APPLIED ELECTRICITY

APPLIED ELECTRICITY

A TEXT-BOOK OF ELECTRICAL ENGINEERING
FOR SECOND YEAR STUDENTS

BY

J. PALEY YORKE

HEAD OF THE PHYSICS AND ELECTRICAL ENGINEERING DEPARTMENT
AT THE LONDON COUNTY COUNCIL SCHOOL OF MARINE
ENGINEERING, POPLAR, LONDON, E.

LONDON
EDWARD ARNOLD
41 & 43, MADDOX STREET, BOND STREET, W.

1906

PREFACE.

THIS book is intended as a Text-book of Electrical Engineering for second year students. I define these as students who have already become acquainted with the elementary fundamental principles and laws of Magnetism and Electricity, and who have also a knowledge of the elements of Mechanics, Heat, and Mathematics. It may seem that I am assuming too much for the average student in a Technical Institute, but if he is to gain any really useful knowledge of this subject such preliminary training is absolutely essential. My experience teaches me that students are quite prepared to devote their first year to this preliminary work when the necessity for and the subsequent advantage of it are pointed out.

This volume is intended to be a direct "follow-on" to this first year work.

In it I have endeavoured to describe the fundamental principles of applied magnetism and electricity, and have been careful to show how the practical applications are related to the phenomena of "pure" science. In fact, I have tried to treat the subject with absolute continuity, so that there shall be no line of demarcation dividing the region of pure from that of applied electricity. But, of course, I have only dealt with those sections of pure science which are connected with modern electrical practice.

Wherever it has been possible I have described laboratory experiments in some detail, so that the book will be useful not only as a supplement to a course of lectures but also in the laboratory, where the student may perform the experiments and make determinations himself.

The scope of the book is intended to be such that the reader who has mastered its contents may proceed profitably to the study of the many excellent specialised works on such branches as electrical machine design, alternate current work, and the like. Indeed, it is intended as a connecting link between elementary magnetism and electricity and these specialised branches of Electrical Engineering.

I hope that the drawings and data of the machines given in the Appendix will prove useful to those lecturers who wish to extend the course to a third year.

I am obliged to Messrs Siemens Bros. for the drawings of the continuous current machine, and to the British Electric Plant Co., Alloa, Scotland, for those of the alternator which was designed by Dr Max Breslauer. In this connection my thanks are also due to Mr W. E. Robson. I am further obliged to Messrs Crompton & Co.; to the Bastian Mercury Vapour Lamp Co.; and to the "W. J. Davy" Arc Lamp Co. for the loan of illustrations of the mechanism of their lamps.

I am indebted to Mr. J. Charlesworth, B.Sc., of the Northern Polytechnic Institute, for valuable assistance in proof reading and in the compilation of data and examples.

J. PALEY YORKE.

LONDON, 1906.

CONTENTS.

CHAPTER I.

INTRODUCTORY.

CHAPTER II.

MEASUREMENT OF CURRENT STRENGTH.

CHAPTER III.

MEASUREMENT OF E.M.F.

CHAPTER VII.

ELECTRIC INSTALLATIONS AND POWER DISTRIBUTION.

CHAPTER VIII.

MAGNETISM.

CHAPTER IX.

ELECTRO-MAGNETISM.

CHAPTER X.

ELECTRO-MAGNETIC INDUCTION.

CHAPTER XI.

DYNAMO PRINCIPLES—ALTERNATOR ARMATURE.

CHAPTER XII.

DYNAMO PRINCIPLES—CONTINUOUS CURRENT ARMATURE.

CHAPTER XIII.

DYNAMO PRINCIPLES—FIELD MAGNETS.

CHAPTER XIV.

CONTINUOUS CURRENT MOTORS.

CHAPTER XV.

ALTERNATING CURRENTS AND TRANSFORMERS.

CHAPTER XVI.

SECONDARY CELLS: ENERGY METERS.

APPENDIX (pp. 393—416).

INDEX (pp. 417—420).

CHAPTER I.

THE UNITS OF MEASUREMENT AND THE RELATIONSHIP BETWEEN THEM.

ELECTRICITY is capable of producing many different effects. The nature of these depends upon circumstances. A "quantity" of electricity may be discharged through a compound liquid and the effects produced would be chemical, heating, and magnetic. The liquid would be decomposed; it would be heated by the passage of the electricity; and about it there would be a magnetic field. But if a similar quantity be discharged through a wire there would only be two effects produced, the heating and the magnetic. Again, if a similar quantity be "at rest" on a body there will be no evidence of any of these effects; but it can be shewn that the body is capable of attracting to itself any light particles of matter, and of repelling other bodies similarly charged with electricity. This charged body would not be heated by the resident electricity; there would be no sign of any magnetic field in its neighbourhood; and if the charge were resident on a liquid there would be no chemical effect as a result. These are elementary facts and are summed up in the statement that the phenomena of electricity in motion are different from those of electricity at rest upon bodies.

When a body is charged with electricity it is capable of doing work—it may attract light particles of matter or repel similarly charged bodies—and it therefore possesses energy in addition to its gravitational energy. This additional energy is electrical energy and depends upon the quantity of electricity and the potential of the body.

P. Y.   1

There is an exact parallel in mechanics. A body possesses mechanical energy in virtue of its quantity of matter and of its position relatively to the earth. If either of these be altered then the energy of the body is altered.

When the quantity of electricity upon a body remains unaltered without external interference the electricity is said to be "at rest" upon the body. If an additional charge be imparted to the body its energy will be increased. If a charge be removed from the body its energy will be decreased. But in order to alter the amount of energy in each case electricity must be put in "motion."

A cistern of water is capable of doing work. Its energy depends upon the quantity of water and upon the level of the water. In order to alter its energy water must be put into motion. Either water must be run out or be run in or the whole cistern must be moved bodily. Now when water is being run out of the cistern the quantity of water in it is decreasing and the level of the water is also decreasing. The outflowing water may be used to turn a water wheel and so do work—but this work is done at the expense of the store of energy in the cistern. If the water is to continue doing the work of wheel driving then clearly the store of energy in the cistern will have to be reinforced. Water will have to be pumped up to it at such a rate that the level remains constant. This will entail the expenditure of energy by some external source. And this is true for all cases of water motion. The ceaseless flow of the rivers is maintained by the expenditure of part of the energy of the sun.

A body charged with electricity can be compared to the cistern of water. The body may be put into electrical communication with the earth. Electricity will "flow" from one to the other—but the transference will entail an alteration in the energy of the body. The electricity may do work, but it is at the expense of the energy of the charged body. If it is desired to keep up a ceaseless "flow" of electricity, then some means must be adopted which will maintain the charged body in a constant state—some means of pumping up electricity to the body in order to keep up the quantity and to keep up its level.

To carry the analogy further, electricity can only be kept "in motion" by the expenditure of energy in some form or other. The primary object of this introduction is to point out the ideas involved in the phrases "electricity at rest" and "electricity in motion." The nature of electricity is unknown and thus one cannot describe the way in which electricity moves. Neither can one state the precise meaning of expressions like a "current of electricity" or a "flow of electricity." One is only too apt to compare the transference of electricity with that of water or gas; and to imagine the "flow" of electricity in the wires accordingly. It is well then to have some sounder basis on which to work; and this is furnished by the consideration of energy. *Electricity may be said to be at rest on a body when the energy of that body remains constant without any external interference.*

Electricity may be said to be in motion between two points when it is necessary to expend external energy in order to maintain the original electrical conditions of those points. For the moment it is matterless how electricity moves; the fact remains that when energy is being expended to keep up the electrical conditions of two points, the electricity is said to be moving. A current of electricity is passing.

Practical Units of Quantity and Current Strength. When a given quantity of electricity is put into motion it can produce a certain amount of an effect. If equal quantities be passed through a number of vessels each containing a solution of a silver salt, there would be deposited equal quantities of silver on each of the kathodes of the vessels. This fact furnishes a mode of definition of a unit quantity of electricity. The practical unit is thus defined as that quantity of electricity which when passed through a solution of silver nitrate (made in accordance with a definite specification) deposits $0·0011181$ grammes of silver. This unit is called a *Coulomb.*

Now the coulomb may be passed in a short time or a long

time, and it is obvious that the rate at which the electricity is transmitted will thus be capable of variation. It is well to note that "rate" means *quantity per second*. This is expressed by the phrase "the strength of the current," or the "current strength." The unit of current strength will be clearly one unit quantity per second, and when one coulomb per second passes any cross section of a conductor the strength of the current in that conductor is said to be *one ampere*. The Board of Trade defines an ampere as "the unvarying electric current which when passed through a solution of nitrate of silver in water, in accordance with the specification marked A, deposits silver at the rate of 0·0011181 grammes per one second."

Absolute definitions of the Coulomb and Ampere. The above definitions are those of the practical units. Both the units and the mode of definition are derived, however, from the "absolute" units which are defined on a mathematical and mechanical basis.

The absolute unit of current strength—on a system known as the electro-magnetic system—is defined as the strength of the current which, flowing through a conductor bent into a circle of unit radius (1 cm.), will exert a force of 2π dynes, on a unit magnetic pole placed at the centre of the circle, the force acting along the axis of the circular coil. This will be discussed fully in Chapter V, but it may be pointed out here that the ampere—the practical unit of current strength—is taken as one-tenth of the electro-magnetic unit defined above. Thus the ampere may be more scientifically defined as that current which flowing through a conductor bent into the form of a circle with a radius of one cm. will produce a force of $\frac{2\pi}{10}$ dynes on a unit magnetic pole placed at the centre of curvature of the conductor.

Potential; Difference of Potential; E.M.F. A body charged with electricity possesses energy in virtue of its charge and of its electrical potential. A charged body may be put under such conditions that its potential is zero, that is to say, is equal to the potential of the earth. The body will be incapable of doing electrical work in such a case—and will therefore have no energy.

Potential may be regarded as that electrical condition of a body which determines whether electricity will flow from it or to it when it is put into electrical communication with the earth, the electrical potential of the earth being called zero. If electricity flows from the body to earth, the potential of the body is positive; if electricity flows to the body from earth, the potential of the body is negative.

This may be compared to the motion of water with respect to sea level. A, B and C in Fig. 1 are three equal cisterns sunk below the level of the sea and provided with taps at the

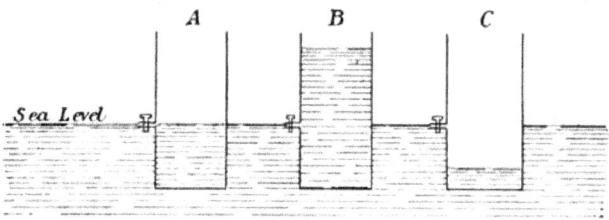

Fig. 1.

surface of the water. Thus the tanks can communicate with the sea. Now if they be filled with water up to sea level they will be equivalent to charged bodies at zero potential. If A's tap be opened there will be no flow of water either to or from the cistern. If B be filled with water to a higher level than the sea and the tap opened, water will immediately flow to the sea. B is equivalent to a charged body at a higher potential than the earth. Work had to be done to raise the water level in the cistern and consequently energy was stored up in it.

If C be now partially emptied, the inside level of water will be lower than the outside. Work is necessary to do this. C therefore possesses energy although it is numerically negative, for when the tap is opened water will flow from the sea to the cistern.   It will be seen that the level of the sea will not be appreciably altered in either case and may be called zero.

The potential of a body with respect to the earth can be directly compared to the level of water with respect to sea level. A difference of water level causes a pressure tending to equalise the levels.   It sets up a force which could be called a water-motive-force.   If the resistance to be overcome is not infinitely great then that force will cause the water to move in the direction in which it is acting.

Similarly a difference of electrical level, that is of potential, causes an electrical pressure tending to equalise the potentials. This electrical pressure tends to urge electricity from the place of high potential to the place of low potential.   This is a force tending to move electricity and it is called an electro-motive-force. If the resistance to be overcome is not infinitely great this electro-motive-force will cause electricity to move in the direction in which it is acting.

If two bodies be at the same potential above or below zero, then if they be put into electrical contact there will be no transference of electricity.   But if they be at different potentials above or below zero, then there will be a difference of potential between them.   Suppose, for example, that a body A has a potential represented numerically by $+100$, that is 100 units above zero, and a body B has a potential $+80$. The difference of potential between these bodies is 20 units. If the potential of B be reduced to say, -80 units, then the difference of potential between them would be 180 units.

Now in cases of water pressure the pressure per sq. inch is directly proportional to the *height* of water: and if two cisterns are in communication the force tending to equalise the levels is proportional to the difference in height.

Similarly the electrical force tending to equalise the potentials of two bodies is directly proportional to the *difference* in potential between the two bodies.   Thus in the above-mentioned cases the electro-motive-force between A

and B would be represented by 20 units and 180 units respectively.

Difference of Potential measured in terms of Work. In order to maintain a constant flow of water it is essential to maintain a constant pressure. This may be done by maintaining a constant difference of level between the water in the reservoir and that of the receiver. This can only be done by the expenditure of work, for water will have to be pumped up into the reservoir at the same rate as it is running out in order to keep the level constant.

Now the difference in level between the receiver and the reservoir could be measured in terms of the work required to pump up a definite quantity of water (say 1 gallon) from the one level to the other. It might be said, for example, that if 1000 units of work were required to raise one gallon of water from one level to another, 1000 would represent numerically that difference in level. And if the quantity of water used be always the same, then the work done would certainly be a measure of the difference of level. But it would be more than that—it would be a means of determining how much work was being done by the reservoir in discharging. For if the quantity discharged in a given time be known, then the product of this quantity and the mean difference of level (expressed in terms of work on a unit quantity) will give the total work done.

This is the basis of the measurement of electrical differences of potential. To maintain a constant flow of electricity against a constant resistance, it is necessary to maintain a constant difference of potential. Electricity has to be metaphorically "pumped up" to the point of higher electrical level at the same rate as it is being discharged to the point of lower level. And to do this, work is necessary—the amount of work depending upon the quantity of electricity required and upon the difference of potential.

Unit of Potential Difference and E.M.F. Difference of electrical potential between any two points is measured by the work required to move a unit quantity of electricity from one point to the other. The practical unit of potential difference is called the *Volt*, and one volt difference of potential is said to exist between two points when 10,000,000 ergs of work* must be done in order to move one coulomb of electricity from one point to the other.

When one volt difference of potential exists between two points the electro-motive-force (E.M.F.) between those points is said to be one volt. It is clear that the force tending to move the electricity will be proportional to the difference of potential, and hence they are measured in the same units.

From the definition of the volt it will be seen that if a current of one ampere is flowing in a circuit the E.M.F. of which is one volt, 10,000,000 ergs of work will be done by the current every second, since one ampere is equivalent to one coulomb per second.

Similarly it can be seen that E volts × C amperes will give the work done every second by a current C amps flowing in a circuit of E.M.F. $= E$ volts.

The practical definition of the volt is as follows : " A volt is the electrical pressure that if steadily applied to a resistance of one ohm will produce a current of one ampere, and which is represented by 0·6974 or $\frac{1000}{1434}$ of the electrical pressure at a temperature of 15° C. between the poles of the voltaic cell known as Clark's cell, set up in accordance with the specification marked B." Clark's cell is described in Chapter III.

Resistance. Matter in motion always experiences resistance to the motion and this is generally known as friction. Electricity in motion also meets with *resistance*, the nature of which cannot be stated. This is obvious when it is remembered that the nature of a current of electricity is not known. At the same time resistance is offered by every electrical circuit depending upon the materials of the circuit,

---

* 1 erg is the work done when a force of one dyne is overcome through the distance of 1 centimetre. 10,000,000 or 10^7 ergs $= 1$ *joule*. 1 joule $= 0·74$ foot lbs.

the length, the area of cross-section of its various parts, and its temperature. The effect of resistance is to determine the strength of a current which can be maintained by a constant E.M.F.

The practical unit of resistance is called the Ohm and is defined by the Board of Trade as "the electrical resistance of a uniform column of mercury 106·3 cms. long and 1 sq. mm. in area of cross-section at a temperature of 0° C." It may also be defined as that resistance which requires an E.M.F. of 1 volt to maintain a current of 1 ampere against it.

Ohm's Law. The relationship between current strength, E.M.F., and resistance is expressed by the law known as Ohm's law (Dr G. S. Ohm, 1826). This law states that the strength of a current is directly proportional to the E.M.F. and inversely proportional to the resistance offered. Thus if C denotes current strength, E electro-motive-force, and R resistance,

$$C \propto E \times \frac{1}{R}.$$

If C, E, and R be expressed in terms of certain units, then the sign of equality may be substituted for that of variation. The ohm has been so chosen that if C be expressed in amperes, E in volts and R in ohms,

$$C = \frac{E}{R}.$$

The law holds good either for a whole circuit or for a part thereof. Fig. 2 is a diagrammatic representation of a simple circuit consisting of a cell of E.M.F. E, and of resistance b;

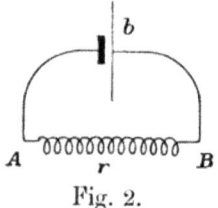

Fig. 2.

and the cell is joined up to a resistance coil AB of resistance r, by connectors whose resistance is negligible.

By Ohm's law the current C in the circuit $= \dfrac{E}{R}$, where $E =$ the E.M.F. and $R =$ the *total* resistance. This total resistance $= b + r$.

Thus
$$C = \frac{E}{b+r}.$$

But the law may be applied to the resistance AB. The current in this resistance will equal the E.M.F. acting between A and B divided by the resistance between A and B.

Let $e =$ the E.M.F. acting between A and B.

Then
$$\text{Current in } AB = \frac{e}{r}.$$

But the current strength is the same in all parts of a simple circuit and therefore this current will be equal to that obtained by considering the whole circuit.

Thus
$$C = \frac{E}{b+r} = \frac{e}{r}.$$

A simple circuit is such that there is only one path for the current to traverse. Fig. 3 is a compound circuit in which

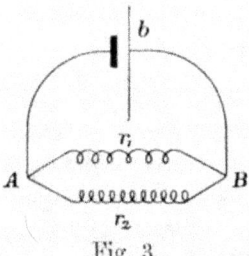

Fig. 3.

the current can split at A, part passing along r_1, and the rest along r_2. The strengths of the currents in these parts would not necessarily be equal. They would be inversely proportional to the resistances of the parts—in accordance with Ohm's law.

To return to consideration of Fig. 2. The difference between E and e is accounted for by the fact that some proportion of the total E.M.F. must be used in transmitting the current through the cell itself.

This will be $E - e$.

Now if $b = r$ then the resistance of the cell—the inside resistance—will be equal to the outside resistance. Therefore there will be as much work done inside as outside. In this case e will be one half of E, and the total work done in one second will be $E \times C$ units. The work done outside will $= e \times C$ units and the work done inside will $= (E - e) \times C$ units, and

$$E \times C = eC + (E - e)\, C.$$

Now if the outside resistance is greater than the inside resistance $E - e$ will become smaller, and when the outside resistance reaches infinity $E - e$ will vanish. The appreciation of this requires an application of Ohm's law from another point of view.

If $C = \dfrac{E}{R}$ then $E = C \times R$, and this will apply equally to a whole circuit or to a portion. In the simple circuit under consideration therefore, since $E = C \times R$, and since R is the total resistance,

$$E = C\,(r + b).$$

Now in the part of the circuit AB—the outside part—the E.M.F. acting between A and $B = e = C \times r$.

In the inside the E.M.F. urging the current through the cell will be equal to the product of the current and resistance of cell.

Thus $\qquad\qquad (E - e) = C \times b.$

Thus $\qquad E = e + (E - e) = Cr + Cb.$

Now if E and b are constant, as they would be for a given case, it is clear that as r increases the total current will decrease—though not in proportion, since b is a constant in the denominator. But as r increases then the value of Cr will increase over the value of Cb. That is to say the E.M.F.

required to urge the current through the resistance AB will become greater in proportion to that required to urge the same current through the cell as the resistance r becomes greater than the resistance b.

This means then that the work done outside will be greater than that done inside—and the circuit will become more efficient as a result. Thus when the outside resistance becomes so great that the inside resistance is negligible in comparison, e will $= E$ for all practical purposes.

This demonstrates well one very important point, that the greatest efficiency is obtained when the greatest proportion of work is done in the outside resistance. And this is obtained when the outside resistance is great when compared to the inside resistance.

The following example illustrates the various cases. Let $E = 2$ volts and $b = 0\cdot1$ ohm. If an external resistance r be connected up, having a resistance (i) $0\cdot1$ ohm, (ii) 1 ohm, (iii) 10 ohms, the following will be obtained :

	(i)	(ii)	(iii)
b	$0\cdot1$ ohm	$0\cdot1$ ohm	$0\cdot1$ ohm
r	$0\cdot1$ ohm	$1\cdot0$ ohm	$10\cdot0$ ohm
C	$10\cdot0$ amps	$1\cdot82$ amps	$0\cdot198$ amps
E	$2\cdot0$ volts	$2\cdot0$ volts	$2\cdot0$ volts
e	$1\cdot0$ volt	$1\cdot82$ volts	$1\cdot98$ volts

These results should be checked by the reader.

The application of Ohm's law in the above case is a frequent cause of trouble to students, but its importance cannot be overestimated. Students are therefore urged to go over the arguments again and again until they feel that they have got a clear mental understanding of the various points involved. They must not be content with merely understanding the simple juggling of the mathematical expression of the law, for such understanding is not of itself sufficiently useful for further application.

A similar line of argument can be pursued with regard to a dynamo. The work done in the outside circuit must be greater than that done inside in order to have an efficient dynamo. Now a dynamo is generally needed to give a *constant* E.M.F. in an outside circuit—say 100 volts. If the outside resistance be altered then the relationship between the outside E.M.F. and that required to urge the current through the dynamo would be altered. Thus the outside E.M.F. would vary unless, either (*a*) the resistance of the dynamo is so small that it is negligible compared with all the possible outside resistances, or (*b*) some means is taken whereby the outside E.M.F. is kept constant although the total E.M.F. may vary—either by increasing the speed of the dynamo or increasing the strength of its magnetic field.

In the case of a cell as the total current increases due to decrease of total resistance, the chemical energy expended in the cell becomes greater—in order to maintain the E.M.F. of the cell. In the case of the dynamo, as the total current increases due to decrease of total resistance more mechanical work must be done by the driving engine to maintain a constant speed and therefore a constant E.M.F.

CHAPTER II.

MEASUREMENT OF CURRENT STRENGTH.

THE various effects of an electric current are proportional in some way to the strength of the current. The definitions of the unit of current strength suggest this, and furnish a basis for the measurement of current strength.

The chemical effect of a current is, according to Faraday's electrolytic laws, directly proportional to the strength of a current and to the time during which the current passes. Moreover the amount of any ion deposited or liberated by unit current in unit time depends upon the chemical nature of that ion. For example 0·0011181 grammes of silver would be deposited by 1 ampere in 1 second—according to the practical definition of an ampere—but in the same time and with the same current there would be deposited :

          0·0003281 grammes of Copper.
          0·000679   ,,   ,, Gold.
          0·000304   ,,   ,, Nickel.
          0 000337   ,,   ,, Zinc.
          0·0000103   ,,   ,, Hydrogen.
          0·0000828   ,,   ,, Oxygen.

These amounts are known as the *Electro-Chemical Equivalents* of the respective substances.

In one hour therefore there would be deposited 3,600 times these amounts by a current of 1 ampere. There would be 4·0252 grammes of silver and 1·1812 grammes of copper.

The Silver Voltameter. This provides one of the most accurate methods of measuring the strength of a current, a method not adopted in everyday practical work but used for the standardising of all kinds of practical current measurers. The method is slow in action and requires much experimental skill. The main apparatus used is called a *voltameter*—not to be confused in any way with a voltmeter—and for most accurate work this consists of a platinum bowl which contains the electrolyte and at the same time acts as the kathode of the voltameter.

The platinum bowl is carefully weighed and into it a solution of silver nitrate, made up in accordance with the Board of Trade specification*, is introduced. A plate of silver is suspended in the liquid by platinum wires in such a manner that the plate is horizontal. The silver plate is wrapped in filter paper as it disintegrates and particles falling on the bowl would cause error. This arrangement constitutes the voltameter. The current to be measured is passed through the voltameter for a known time, the silver plate being made the anode and the weighed bowl the kathode. At the end of the allotted time—which must be accurately taken—the current is switched off. The solution is poured out of the bowl, which is thoroughly washed with distilled water. It is then dried—not with a cloth!—in a drying oven, and when at the temperature of the air is again weighed. The increase in weight is the amount of silver deposited by the current in the known time. Knowing the electro-chemical equivalent of the silver the strength of the current which was passing through the voltameter is easily calculated. If this current has not been constant in strength the result obtained will be the strength of the *average* current passing through the voltameter during the time chosen.

It is readily seen that this method whilst admitting of great accuracy is too slow for everyday work, and of no value

* "The liquid should consist of a neutral solution of pure silver nitrate, containing about 15 parts by weight of the nitrate to 85 parts of the water."

for the indication of currents of continually varying strength. Its real value lies in its use for the standardisation of instruments working on other principles, and an application of this is given below.

The Tangent Galvanometer. A galvanometer is an instrument which may be used for detecting or measuring currents of electricity, depending for its action on the fact that about a wire conveying a current there is a magnetic

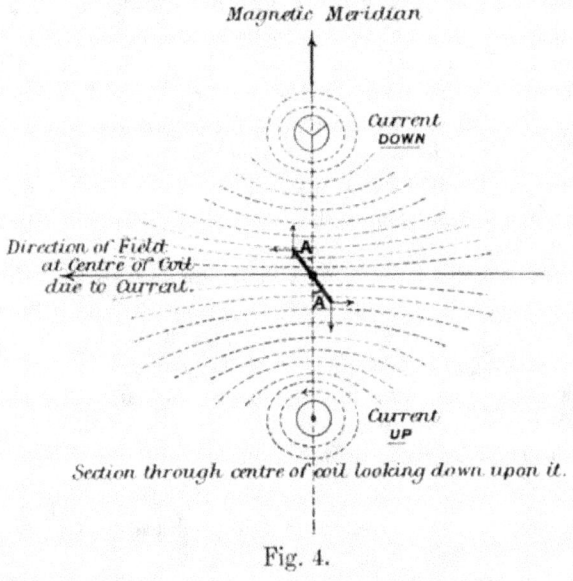

Fig. 4.

field, the strength of which is in direct proportion to the strength of the current.

The tangent galvanometer is perhaps the simplest form, consisting of a thin coil of wire in a vertical plane and having pivoted at its centre a small magnetic needle capable of moving horizontally about its pivot. This needle is provided with a long light pointer at right angles to its length which moves over a scale graduated in geometrical degrees.

When the galvanometer is to be used it is arranged that the coil shall be in the magnetic meridian and thus the needle will lie in the plane of the coil, and the pointer will be over the zero of the scale. The needle is being acted upon by a couple, due to the earth's magnetic field, tending to set it in the earth's magnetic meridian. This couple is called the controlling couple.

Now when a current is passed through the wire of the coil there will be set up a magnetic field about the coil. Near the centre of the coil this field will be uniform and will have a direction at right angles to the plane of the coil and therefore to the magnetic meridian. The needle is small in order that it shall be always in this uniform part of the field acting at right angles to the plane of the coil and along its axis. Thus there is acting on the needle a couple—called the deflecting couple—which is endeavouring to set the needle in a direction at right angles to the plane of the coil. Fig. 4 illustrates this.

Whenever two couples act at right angles to each other then the tangent of the angle of deflexion produced is proportional to the deflecting force. Thus in the case of Fig. 4 the tangent of the angle A is proportional to the strength of the magnetic field set up by the current. This in turn is proportional to the strength of the current. Thus if A represents the angle of deflexion and C the strength of the current,

tangent of A $\propto$ strength of field produced by C.

$$\therefore \tan A \propto C.$$

But if the angle of deflexion varies with C in such a way that its tangent is proportional to C it follows that

$$\tan A \times \text{some number} = C.$$

This number when obtained is called the *constant* of the galvanometer and may be denoted by K.

Thus $\qquad \tan A \times K = C.$

Hence a tangent galvanometer whose constant is known
can be used as a current measurer. The angle of deflexion
produced by any current is noted, the tangent of that angle
is found by reference to a table of tangents (see Appendix)
and this tangent is multiplied by the constant. The product
is the strength of the current.

**Determination of the Electrolytic Constant of a
Tangent Galvanometer.** The constant may be determined
by measuring a current which produces a certain deflexion,
using for the purpose a voltameter. The galvanometer and
the voltameter are connected up in series with each other
and with a battery of secondary cells—to ensure a constant
current—a variable resistance and a switch as shewn in Fig. 5.
The resistance is used to adjust the current at the outset in
a preliminary trial. In making this the resistance should be
adjusted so that the deflexion produced of the galvanometer
needle does not exceed 60°. If the deflexion be greater there

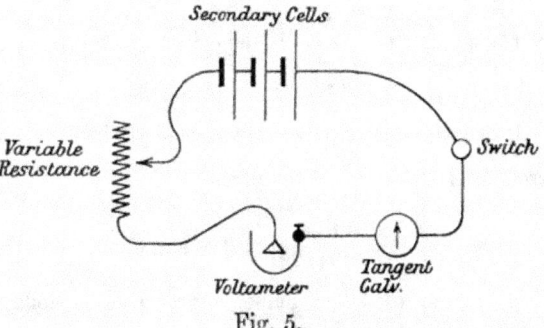

Fig. 5.

is a greater liability of error in the final result as the tangents
of angles over 60° increase out of all proportion to the increase
in the angle. A glance at a table of tangents will shew this,
and it will be seen that a small error in reading at the higher
part of the scale will produce a relatively large error in the
result.

This is the result of an experiment,

Mass of kathode at start = 51·3 grammes.

Time of switching on current = 8.15 p.m.

Time of switching off current = 8.35 p.m. to $\frac{1}{2}$ a second.

Deflexion produced in galvanometer = 42·5°.

This was noted every minute to see that current was constant.

Mass of kathode at end of experiment = 52·76 grammes.

From this data it follows that

the mass of silver deposited = 1·46 grammes,

the number of seconds during which the current was passed = 1200.

Now 0·0011181 gms. of silver are deposited in 1 sec. by a current of 1 ampere.

∴ 0·0011181 × 1200 grammes of silver are deposited in 1200 secs. by a current of 1 ampere.

∴ 1·34172 grammes of silver are deposited in 1200 secs. by a current of 1 ampere.

∴ 1 gramme of silver is deposited in 1200 secs. by a current of $\dfrac{1}{1\cdot34172}$ ampere.

∴ 1·46 grammes of silver are deposited in 1200 secs. by a current of $1\cdot46 \times \dfrac{1}{1\cdot34172}$ amperes.

$$\therefore C = \frac{1\cdot46}{1\cdot34172} = 1\cdot088 \text{ amperes.}$$

This current of 1·088 amperes produced a deflexion of 42·5° with the galvanometer used.

$$\therefore \tan 42\cdot5 \times K = 1\cdot088,$$
$$0\cdot9163 \times K = 1\cdot088.$$

Whence $$K = \frac{1\cdot088}{0\cdot9163} = 1\cdot19.$$

Hence the constant of the particular galvanometer used is 1·61 for amperes, and this holds good for all places where the horizontal component of the earth's magnetic field is the same as at the place where the determination was carried out.

Reflecting Galvanometers. Other galvanometers may be similarly standardised. For example, a mirror galvanometer is used with a lamp and a translucent scale. A beam of light acts as the pointer and moves across the scale when the needle, to which the mirror is affixed, is deflected. Now this pointer of light can be made any length by adjusting the source of light and the scale at various distances from the galvanometer. In this way a given deflexion of the mirror will produce a greater displacement of the beam of light along the scale as the scale is further away from the needle. And a further advantage of the beam of light is that the angle through which it is turned is twice the angle through which the needle is turned.

With mirror galvanometers used under these conditions it is always possible to work with very small deflexions of the needle, with the result that the distances along the scale become proportional to the tangents of the angles of deflexion of the needle. Hence it is usually taken that with mirror galvanometers the deflexion—that is the distance which the spot of light is moved along the scale—is proportional to the current. But it should be remembered that this is so because those deflexions are proportional to the tangents of their several angles, and also to the *small* angles themselves.

And this holds good for all kinds of mirror galvanometers—provided that the deflexions are not too great compared with distance of the scale from the needle. The distance generally chosen for laboratory and test room work is 40 inches.

To standardise a mirror galvanometer the current required to produce one division of the scale deflexion at a scale distance of 40 inches is determined—by experiment and calculation. This is generally extremely small and could not be measured directly by means of a voltameter. The maker of a good instrument generally supplies a certificate of constant, of which the following is a copy.

Figure of Merit of Galvanometer No. 19172.

This galvanometer gives 1 scale division ($\frac{1}{40}$th inch) deflexion with 1 volt on **1020** Megohms at a distance of 40 inches from the scale.

Resistance 870 ohms at 17° C.

Focal length of mirror 40 inches.

From this it is seen that in the case of the galvanometer quoted a current of $\dfrac{1}{1,020,000,000}$ ampere will produce a deflexion of one scale division ($\frac{1}{40}$th inch) at a scale distance of 40 inches. Thus if a deflexion of 150 divisions is produced, the current producing it must be $150 \times \dfrac{1}{1,020,000,000}$ ampere.

Kelvin Astatic Galvanometer. Kelvin's astatic galvanometer is generally one of very high resistance and is

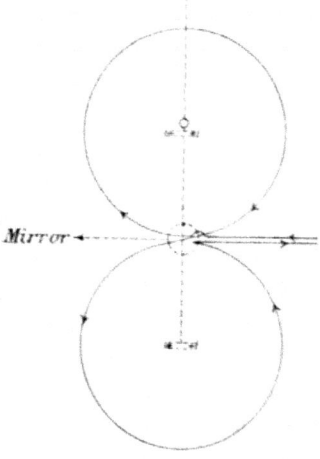

Fig. 6.

consequently particularly adapted for high resistance measurements and for all purposes for which the galvanometer is used as a voltmeter. The needle of the galvanometer is an astatic pair of small magnets, and each magnet of the needle is hung in the centre of a coil, as illustrated diagramatically by Fig. 6. These coils may be connected in series or in parallel according to the resistance required, but they must be so connected that the magnetic fields in each tend to deflect the needle in the same direction. If they produced opposite effects there would be no deflexion provided that the effects of each coil were equal.

The needle is suspended by a fine fibre of unspun silk and a mirror is attached, as shewn in the figure.

If the needle were astatic, the earth's field would have no effect upon it; but it is practically impossible to have a perfectly astatic pair. However the controlling forces are provided by means of a magnet at the base of the instrument: and these can be weakened or strengthened at will by means of a pair of magnets at the top of the case. These magnets are capable of adjustment—they can be made to help or oppose one another, and to decrease or increase the effect of the controlling magnet. Thus various degrees of sensitiveness may be obtained.

This particular form of galvanometer is usually wound with a resistance of some 20,000 ohms, and in its condition of maximum sensitiveness may indicate a current of some $\dfrac{1}{30,000 \times 10^6}$ ampere, which is ·00003 micro-ampere, a micro-ampere being one-millionth of an ampere.

Ballistic Galvanometer. For some electrical work, such as capacity determination and inductance measurements, a special form of instrument, called a *ballistic galvanometer*, is required. This differs from an ordinary galvanometer in that the needle is a heavy one with a long period of oscillation.

If an electric discharge or a current of brief duration be passed through a galvanometer, then a deflecting couple is set up by its magnetic field. It can be shewn that if the discharge is concluded before the needle of the instrument has been appreciably displaced from its normal position, the effect of the whole discharge will be shewn by the ultimate total displacement and that the *sine of half of the angle of such displacement* is directly proportional to the *total quantity* of electricity discharged.

If the displacement be small it may be said that the *actual deflexion* is *directly proportional to the quantity* discharged. These conditions are realised when the galvanometer is provided with a mirror and scale in the usual way, and the galvanometer can then be used for comparing or measuring (after calibration) *quantities* of electricity. It is the full extent of the first *kick* of the needle which must be taken for the

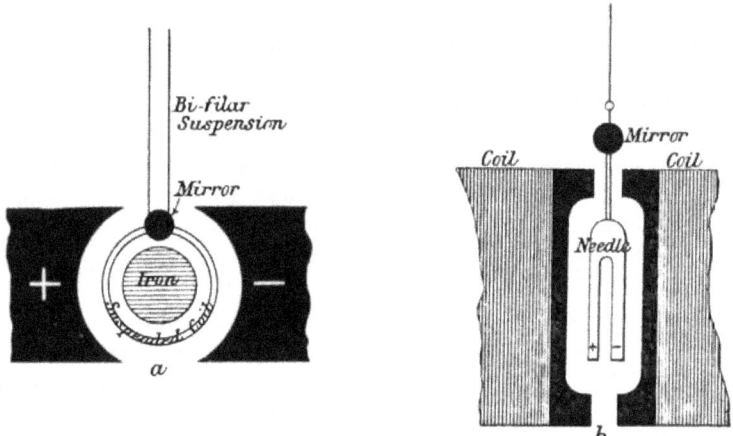

Fig. 7.

actual deflexion. There will not be any permanent deflexion since the discharge was passed before the needle had been sensibly displaced.

Ballistic galvanometers are made either with a suspended coil or with a suspended magnet. In both cases these are comparatively heavy, and the period of oscillation is lengthened by having a long suspension.

For capacity determinations the suspended coil type is

perhaps the better, but for small induction effects the magnet needle yields the more accurate results.

Fig. 7 illustrates the two forms of needles in ballistic galvanometers. *a* shews the coil suspended between opposite poles of a permanent magnet. The coil is hung by two phosphor-bronze strips, the suspension being called bifilar. These strips conduct the discharge to and from the coil. The controlling couple is furnished by the weight of the coil, for when it is deflected it will have to be raised slightly by virtue of the bifilar suspension.

b shews the magnet needle, usually made in the horseshoe or bell form, illustrated. The magnetic axis of the needle is made to coincide with the axis of two coils on either side, when the instrument is to be used. The controlling forces are furnished by the earth's magnetic field augmented or diminished by that due to a magnet placed under the base of the instrument or in some other convenient position.

The uses to which a ballistic galvanometer may be put are described in Chapter X.

The Figure of Merit takes the following form :

Figure of Merit of Galvanometer No. **5902.**

This galvanometer gives 1 scale division ($\frac{1}{40}$ inch) deflexion with 0·0009936 microcoulomb at a distance of 40 inches from the scale with a period of oscillation of **10** seconds.

Resistance 6733·2 ohms at 17° C.

Focal length of mirror 40 inches.

PRACTICAL INSTRUMENTS.

Galvanometers are only for use in measuring extremely small currents, where great accuracy and sensitiveness are required. Moreover they are not direct reading. Whilst these instruments could be used for the reading of larger currents, they would provide a greater degree of accuracy than is required for everyday work and would always require the attention of someone more or less skilled in the manipulation of delicate instruments.

In practical work the strength of a current is measured by an instrument which indicates with a pointer moving over a scale the current strength in amperes. These practical instruments are called *ammeters*—a contraction of ampere-meters—and in use are introduced in series into the circuit in which it is desired to measure the current.

Conditions of a good Ammeter. A good ammeter should fulfil the following conditions :

(*a*) It must measure currents.

(*b*) It must be sufficiently sensitive for the degree of accuracy required. An error of one quarter of an ampere in a reading of 100 amperes might be near enough for all practical purposes ; but an error of a quarter of an ampere in a current of one ampere would be too great even for practical purposes.

(*c*) It must carry the maximum current for which it is designed without undue heating.

(*d*) It should be "dead-beat," so that the pointer would come to rest quickly at its proper indication and not oscillate. If a pointer took an appreciable time to become steady the current might vary during the time, and the final reading obtained would not be the reading which was desired at the outset. And with some circuits the needle might never take up a position of rest.

(*e*) It should not be affected by external magnetic fields. If it were it would be of little use in an engine room, or in proximity to other magnetic instruments.

(*f*) It should have such a low resistance that its inclusion in a circuit would not appreciably alter the total resistance of the circuit.

This last condition is of the greatest importance. To measure a current in a circuit the ammeter must be introduced into the circuit, in series, so that the whole of the current to be measured shall pass through the instrument. If the ammeter has an appreciable resistance compared with that of the circuit, then it is clear that the total resistance of that circuit will be increased by the inclusion of the ammeter. Therefore the current in the circuit will be decreased, and the current indicated by the ammeter will be lower than that which it was intended to measure. Moreover an ammeter with appreciable resistance would absorb an appreciable amount of energy.

Every instrument must of course be acted upon by two sets of forces—one set tending to keep the needle in one position (the zero position), and the other set tending to pull the needle into some other position. The actual position will depend in some way upon the relative values of these two sets of forces. The set tending to keep the needle at zero is known as the controlling couple, its effect being termed the *controlling torque*. The effect of the set tending to pull the needle into some other position is known as the *deflecting torque*.

When the needle of an instrument comes to a position of rest these two torques must be equal for that position of the needle, for the controlling torque represents the tendency of the controlling forces to bring the needle back to zero, and the deflecting torque represents the tendency of the deflecting forces to turn the needle into some other position—parallel to their direction. The position which the needle will take up under these torques clearly represents a position where the two tendencies are balanced.

In many ammeters the controlling torque is due to the weight of the needle—to the fact that the centre of gravity of a mass tends to be vertically under the point of support of the mass. In others the controlling torque is produced by means of a spring.

The deflecting torque is produced by the action of two magnetic fields in the case of all electro-magnetic instruments.

Types of Ammeters. Ammeters may be divided into two types, *electro-magnetic* and *thermal*. And in turn the electro-magnetic instruments may be divided into two; "A" type consisting of an iron needle and a fixed coil, and "B" type consisting of a movable coil (which acts as the needle) and a fixed permanent magnet.

The electro-magnetic instruments depend for their action on the effect of the magnetic field about a coil on an iron needle or on a fixed magnet, according to the type. The thermal instruments depend upon the fact that a current always generates heat in a wire, the heat being proportional to the square of the current. Thus a wire through which a current is passed has its temperature raised. This produces expansion of the wire and the expansion is used to indicate the strength of the current passing.

Kelvin Ammeter. An idea of the Kelvin ammeter, or "ampere gauge," may be obtained from Fig. 8. A solenoid of thick bar copper S is arranged with its axis vertical so that it can attract an iron rod IR into it. This rod is mounted with a swivel mount on one end of an aluminium balance beam B, and it is weighted with a weight W so that it always assumes a vertical position whatever the inclination of the beam may be. The beam can turn about a hook H and at the opposite end to that at which the rod is mounted a counterpoise K is hung. An aluminium pointer P is fixed to the beam and its free end can move a scale as shewn. The counterweight K is such that the pointer lies over the zero on the scale when there is no current in the solenoid. When a current is passed through the solenoid, the iron rod will be drawn up and the pointer will move over the scale. In the case illustrated the ammeter can measure from 50 to 500 amperes.

The instrument is rendered dead beat by means of a piston and dash pot. This consists of a glass cylinder of oil in which a disc of ivory pierced with small holes can slide as a piston slides in a cylinder. This disc is fixed to a continuation of the iron rod below the weight W, so that the dash pot is behind the scale. It is shewn by the dotted lines. The instrument is not a portable one.

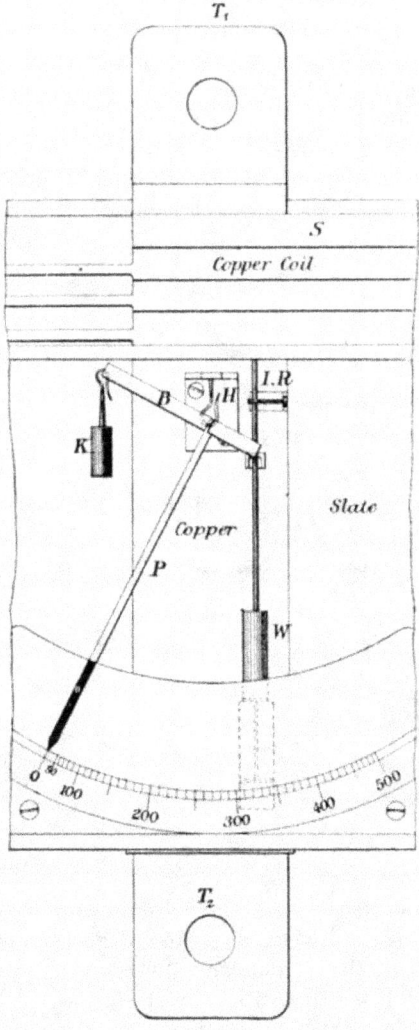

Fig. 8.

A Simple Iron Needle form. A simple form of iron needle instrument is illustrated by Fig. 9. A solenoid S is capable of drawing into it a piece of iron I which is suspended from an arm AO of a balance beam arrangement AOW by means of a thread T. O is the turning point of the beam and a long pointer P is fixed to the beam at this point. An adjusting weight is screwed on at the end W and this is used to adjust the zero position of the pointer.

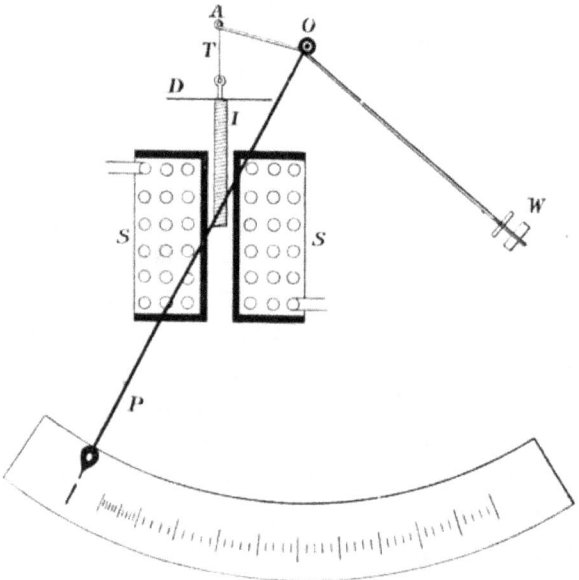

Fig. 9.

When a current passes through the wire of the solenoid the iron will tend to be drawn in to a degree depending upon the strength of the current and the resulting magnetisation of the iron. If it is drawn in, then the pointer P must move over the scale—which is graduated in amperes by original comparison with a standard instrument. Of course when the first instrument of a type has been calibrated in this way, the subsequent instruments are made exactly after the first model—scale included.

In the instrument under discussion there is a thin flat iron plate D arranged horizontally at the top of the iron I. This plate serves two purposes. Firstly it increases the pull due to the magnetic field of the solenoid, and secondly it helps to make the needle dead beat.

This instrument can only be used vertically, which statement applies generally to instruments with a "gravity control."

Disadvantages of Iron Needles. There are many serious disadvantages in connexion with iron-needle instruments. These are chiefly magnetic and are brought about by the fact that the degree of magnetisation of a piece of iron is a varying quantity until the degree of *saturation* is reached, and that this variation is by no means a uniform one. Hence the scale of an ammeter having an iron needle should not be calibrated for any currents which do not produce in the coil a magnetic field strong enough to magnetise the iron to saturation. The mechanical effect on the needle is proportional to the strength of the magnetic field in the coil and to the degree of magnetisation of the needle. The field in the coil will vary proportionately with the strength of the current flowing, but the degree of magnetisation of the needle will vary irregularly with the strength of the field magnetising it. Hence the difficulty of calibrating.

But when the iron has reached its saturation point then it will be unaffected by further increase in the magnetic field : and then the mechanical effect upon it will be in direct proportion to the strength of the magnetic field and therefore to the strength of the current.

Thus it is seen that an iron needle type of instrument is not adapted to the measurement of small currents. It may be conceded that the strength of the magnetic field set up by a small current could be increased by having a large number of turns of wire on the coil; but in such a case the resistance of the ammeter would be increased and one of the most important conditions of an ammeter would be unfulfilled.

Another serious disadvantage lies in the fact that when saturation is not reached the readings going up the scale will not correspond to those coming down for equal currents. This is due to the tendency of all iron to retain its previous magnetic state (Chapter VIII).   The result is that on the downward readings the needle is magnetised to a higher degree than it was for corresponding upward currents. Hence the downward readings are found to be higher than corresponding upward readings.   This will not be found in cases where the iron needle is saturated from the outset of the scale readings : but is common enough in very cheap ammeters.

Comparison of an Ammeter with a Standard. This may be proved very simply by connecting a good ammeter in

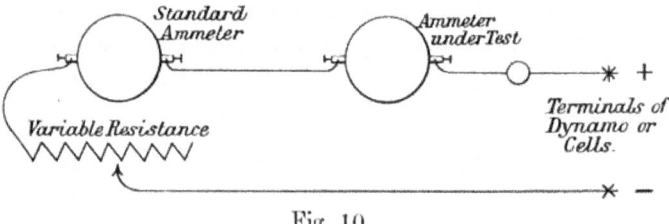

Fig. 10.

series with a cheap iron needle ammeter and a variable resistance as shewn in Fig. 10.   By starting with all the resistance in circuit—say 100 ohms—and gradually decreasing it, the current can be gradually increased to the maximum current for which the instruments are calibrated.   (The resistance coil used should be capable therefore of carrying this current safely without undue heating. For example a resistance box must *not* be used !)

Simultaneous readings should be taken of the instruments, firstly going up the scale and secondly coming down.   The following are results obtained in this way :

Going up the scale			Coming down the scale		
Indications of Standard	Indications of Ammeter under test	Error	Indications of Standard	Indications of Ammeter under test	Error
0	0	0	7·0 amps.	6·95	+0·05
0·9 amps.	0·75	+0·15	5·3 ,,	5·4	−0·1
1·4 ,,	1·3	+0·1	4·4 ,,	4·55	−0·15
2·1 ,,	1·95	+0·05	3·0 ,,	3·1	−0·1
3·0 ,,	3·0	0	2·1 ,,	2·1	0
4·4 ,,	4·45	−0·05	1·4 ,,	1·35	+0·05
5·3 ,,	5·3	0	0·9 ,,	0·8	+0·1
7·0 ,,	6·95	+0·05	0 ,,	0	0

The differences in the readings are best illustrated graphically by means of a curve known as a *curve of differences*, or of *errors*. Along a line *A B* (Fig. 11) distances are marked off

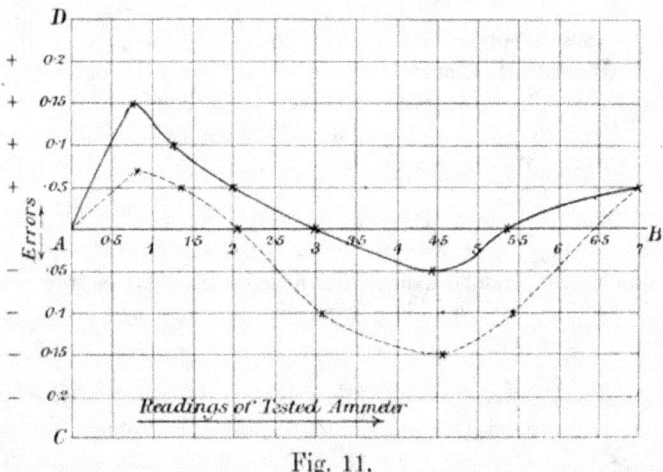

Fig. 11.

from *A*, representing the scale of the ammeter under test. This is easily done on squared paper. The difference between the reading of the tested ammeter and that of the standard ammeter are marked off as ordinates along *D A C*—at right angles to AB. If the tested ammeter reads lower than it ought to read then the difference is marked off above the line AB; and if it reads higher than it should the difference is marked off below the line.

Fig. 11 shews the curve of differences or of errors according to the results given above. The full line shews the errors going up and the dotted line shews the errors coming down the scale. It will be seen that in all cases the readings coming down the scale are slightly higher than those going up. One can only conclude that the iron was not saturated.

This method may be employed equally well for the comparison of instruments in general use with a standard instrument. If a curve of errors be drawn up then the tested instrument may be used to measure currents accurately, for having taken the reading of the instrument one consults the curve—finds from it if there is any error at that reading and if so how much, and corrects accordingly. For example, with the instrument whose curve of errors is given above : if it reads 1·95 with a current whose strength is to be determined one consults the curve, notes that the error is + 0·05 and knows therefore that the true current strength is 1·95 + 0·05 amperes, viz. 2·0 amps. Similarly if it reads 4·45 then the true current strength is 4·45 – 0·05, viz. 4·4 amps.

The instrument can thus be used as a standard for the time being. The drawing up of such a curve of errors is known as "standardising" the instrument. All instruments in everyday use should be standardised from time to time.

COIL TYPES.

The second type of electro-magnetic instruments—type " B "—are those having a movable coil and a fixed permanent magnet. They are, in effect, modifications of the "suspended coil" galvanometer such as the D'Arsonval. These are very reliable instruments as a general rule, and the fact that there is no iron in their constitution prevents their having the same disadvantages as the iron needle type. The coil type of ammeter is in reality a "shunted" voltmeter, and is described in Chapter III.

The Electro-Dynamometer. An instrument used largely as a secondary standard is that known as Siemens's

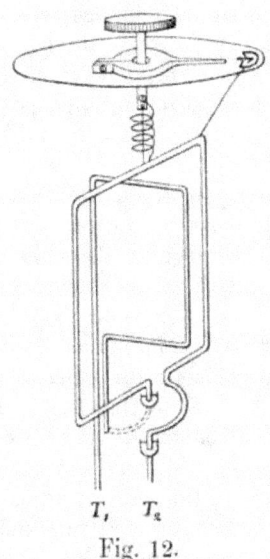

Fig. 12.

dynamometer. In the instrument there are two coils of low resistance and of rectangular form arranged so that their planes are vertical and perpendicular to each other. These coils are arranged so that one is inside the other as shewn in the diagrammatic figure 12. The outer coil is suspended by a silk thread and controlled by a fine helical spring, one end of which is fixed to the coil whilst the other end is fixed to the "torsion head." Electrical connexion to this outer coil—known as the movable coil—can be made by means of mercury cups at the base of the instrument. The ends of the coil dip into these cups, which are placed vertically below the point of suspension of the coil, as shewn by T_1 and T_2 in Fig. 12.

The "torsion" head is an ebonite thumbscrew arrangement which can be twisted round in a hole in the top of the
case of the instrument. The coil being suspended from this
torsion head will be turned one way or the other as the head is
twisted round. To the torsion head a pointer is affixed, which
moves over a scale graduated in degrees at the top of the
instrument. In this way the angle through which the torsion
head is twisted can be measured and this angle gives a measure
of the twisting force applied to the moving coil—a force which
is produced by the twisting of the helical spring about the
suspension thread.

In addition to the pointer on the torsion head there is
another affixed to the movable coil and bent round so that its
end comes over on to the degree scale at the top. It generally
passes through a small slot at the outside of the scale and at
the zero. This prevents the movable coil from being twisted
more than a few degrees from its normal position.

The "fixed" coil, which is the inner coil, is connected in
series with the movable coil so that the current in each coil
shall be the same. The connexions are shewn on the diagrammatic figure.

To use the instrument the coils are adjusted to be at right
angles to each other and at the same time the two pointers are
adjusted to be on the zero of the scale. This can be done by
loosening the pointer grip and turning the torsion head in the
required direction. The pointer can be kept stationary at
zero, and when the coils have been adjusted so that they
are at right angles to each other the pointer on the movable
coil will be at the zero. Then the torsion head pointer is
tightened up and the instrument is ready for use. When the
instrument is received it is generally in proper adjustment and
it is only necessary to adopt the above methods in case of
a renewal of the suspension thread or in case of physical
deterioration of the spring.

Now when the current to be measured is passed through
the dynamometer there will be a magnetic field set up about
each coil. Fig. 13 illustrates the nature of this field. MM'

3—2

represents the movable and *FF'* the fixed coil as seen from above. The directions of the current in each coil and the directions of the resultant magnetic lines of force are shewn. It is seen that the limbs *M'* and *F'* and that the limbs *M* and *F* will tend to approach each other. In short the coils will *tend* to arrange themselves so that their magnetic fields are coincident—so that they are parallel to each other.

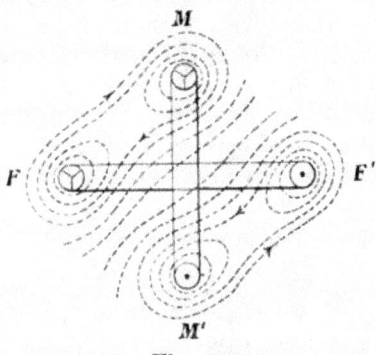

Fig. 13.

Now the pointer of the movable coil passing through a small slot will restrict the movements of the free coil, but at the same time it will move away from the zero and probably to the limit allowed by the slot. The torsion head is now twisted in the opposite direction and to such an extent that the movable coil is twisted back again to its first position with its pointer on the zero.

When this is done then it is clear that the tendency of the movable coil to set itself parallel with the fixed coil has been balanced by the torsion of the spring, tending to twist it to some other position. Therefore the two tendencies are equal.

Now the torque due to the magnetic fields of the currents in the two coils is directly proportional to the strengths of those magnetic fields. And the torque due to the twisting of the spring balancing the deflecting torque is directly proportional to the *angle of torsion*—that is, the angle through which the torsion head had to be turned in order to restore the coil to its original position of equilibrium. It is a well-known law of elasticity that the force producing a twist on a wire or spring is proportional to the angle of twist or torsion produced.

Hence it follows that the angle of torsion (required to balance the deflecting forces) gives a measure of those forces and is proportional to them. Thus

The angle of torsion varies as (the strength of magnetic field in coil 1) × (the strength of magnetic field in coil 2).

But the strengths of these magnetic fields are proportional to the strengths of the currents producing them.

∴ The angle of torsion varies as (the strength of current in coil 1) × (the strength of current in coil 2).

But the coils are in series, therefore the current in each coil is the same. Let it equal C.

∴ The angle of torsion varies as $C \times C$.

I.e.   Angle of torsion $\propto C^2$.

Therefore $\sqrt{\text{angle of torsion}} \propto C$.

Whence it follows that since the square root of the reading is proportional to the current, it must be equal to it when multiplied by some number to be determined. This number is called the constant of the dynamometer.

Hence $\sqrt{\text{angle of torsion}} \times$ a constant $= C$.

The makers supply the constant with each instrument sent out, but it can always be determined by the electrolytic method described at the beginning of the chapter. One has only to know the strength of a current and the corresponding square root of the angle of torsion and the constant may be calculated.

There are generally two ranges of measurement on each dynamometer. These are given by having two sets of windings on the fixed coil—one of a very small number of turns and the other of a larger number. The latter coil is used for the smaller and the former for the greater currents.

The dynamometer is used as a secondary standard, for it has many advantages over the general run of current measurers.

There is no iron whatever in its construction, and it may be used for continuous or alternating currents with the same constant. Its constant may vary after a time by reason of physical deterioration of the spring, but the fresh constant is easily determined and there is no need to calibrate its whole range.

Its only disadvantage is that it is affected by external magnetic fields, and as a result it must be used in a protected place. It is advisable always to set it up so that the plane of the movable coil is at right angles to the earth's magnetic meridian, since any vertical coil through which a current is passing will tend to set itself so that it is at right angles to the magnetic meridian at a place with the lines of force passing through its centre parallel to and in the same direction as the earth's lines. It is not, of course, direct reading, but a table of currents corresponding to the angles of torsion is supplied by the makers.

Hot wire Ammeter. A type of ammeter which indicates the strength of a current by the increase of temperature produced and the consequent increase in length of a wire is illustrated diagrammatically by Fig. 14. ABC is a fine platinum wire fixed at A and B and is taut at normal temperature. AB is "shunted" across a low resistance KL through which the major portion of the main current to be measured passes. However as this resistance KL is constant the fraction of the total current which is "shunted" (see page 69) through AB will be proportional to the whole current.

When AB is heated slightly by the passage of a current it will increase in length and sag. This sag is taken up by another wire CDE fixed to AB at C, and to the base of the instrument at E. Another wire DFG is fixed to CE at D, and passes round a small pulley wheel which can turn about a horizontal axis at F. The end G of DG is fixed to a piece of clockspring HG which keeps it taut constantly. A pointer is fixed on the axis of the pulley F and the indicating end of this can move over a scale as shewn.

When the wire AB is heated it sags, and this sag is taken up by CE and ultimately by the spring HG. In doing this the pulley at F will be rotated somewhat, to an extent depending upon the sag taken up. Thus the pointer will move over the scale, which is calibrated in amperes.

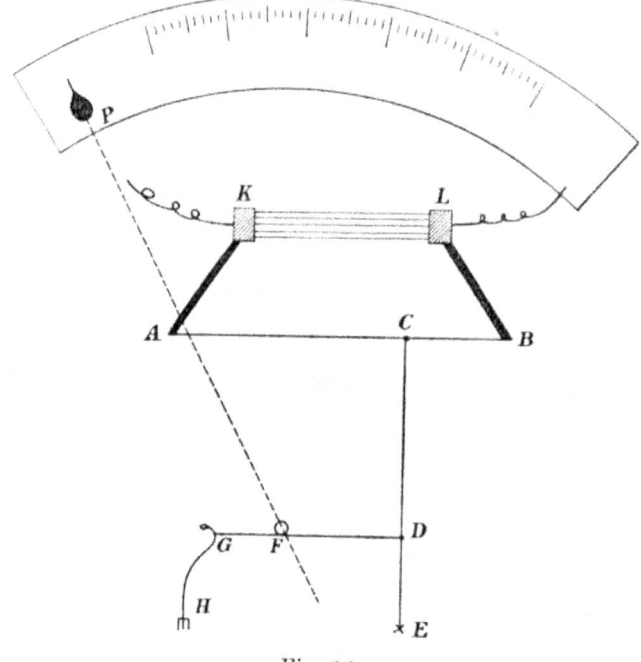

Fig. 14.

The object of arranging the wires as shewn is to multiply the effect of the expansion of AB, depending upon the fact that the sag is greater than the longitudinal expansion.

The instrument is rendered dead-beat by having a light disc of aluminium on the axis of the pulley arranged so that it moves between the opposite poles of a permanent magnet. When the disc moves induced currents are set up in it, and the directions of these currents will be such that their magnetic effects tend to stop the motion producing them.

The hot wire type of ammeter can be used for direct or alternating currents, and it is unaffected by any magnetic fields.

Kelvin Ampere Balance. The standard laboratory instrument for the measurement of current is that known as the Kelvin ampere balance. This belongs to the electro-magnetic group of instruments, and it resembles the dynamo-meter in that there is no iron in its composition and that there are fixed and movable coils.

The movable coils are arranged with their planes horizontal, at the ends of a "balance beam" construction. Parallel to each movable coil a fixed coil is placed above and below, after the manner suggested by the sketch in Fig. 15. F_1 to F_4 represent the fixed and M_1 and M_2 the movable coils. A is

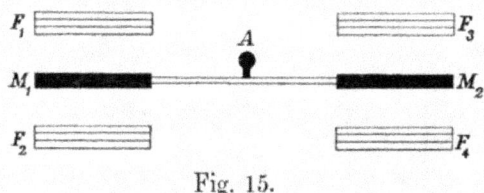

Fig. 15.

the axis about which the "beam" turns. The coils are so connected that when a current is passed through them the combined magnetic effects tilt the beam, M_1 being pulled downwards, say, and M_2 upwards. A sliding weight can be pulled along the beam until equilibrium is restored, and the current strength is determined by reading off the position of the weight upon the beam and multiplying by the constant provided.

This does not intend to be more than the merest outline of the working of the balance. The instrument is used principally for standardising ammeters and dynamometers in general use, but it can only be used by a skilled experi-mentalist. The makers supply the constants, and with them a full description of the instrument and the methods of using it.

CHAPTER III.

ELECTRO-MOTIVE-FORCE AND ITS MEASUREMENT.

THE practical standard of E.M.F. is a cell known as the Clark standard cell. This has an E.M.F. of 1·434 volts at the normal temperature of 15° C.

A section of the cell is shewn in Fig. 16. The metals of the cell are mercury and zinc. The mercury lies at the

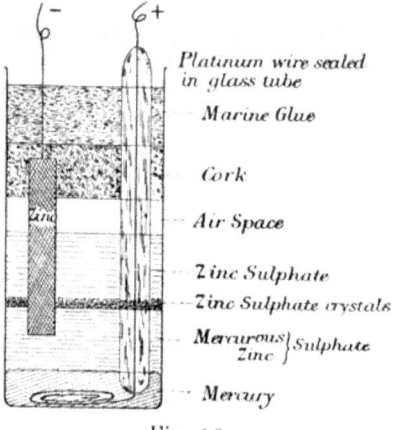

Platinum wire sealed in glass tube

Marine Glue

Cork

Air Space

Zinc Sulphate

Zinc Sulphate crystals

Mercurous} Sulphate
Zinc

Mercury

Fig. 16.

bottom and on this floats a paste of mercurous sulphate and zinc sulphate. Above the paste there is a layer of saturated solution of zinc sulphate—saturation being further ensured by some crystals of zinc sulphate which lie at the bottom of the liquid and on top of the paste.   Above this solution is an air space and then the cell is sealed up with cork and marine glue.

The pure zinc rod passes through the cork down to the middle of the mercurous and zinc sulphate paste.   The upper end of the zinc is connected to one terminal of the case containing the cell and the mercury is connected to the other terminal.   This is done by means of a platinum wire which is sealed in a glass tube as shewn in the figure.   The lower end dips well under the mercury and the projecting platinum is coiled into a flat helix at the bottom of the mercury.

The Board of Trade has drawn up a specification giving details of the construction of the cell and of the materials which must be used.   All cells made up according to this specification are found to have the same E.M.F. as that quoted.

The variation of the E.M.F. of the cell with its temperature has been determined and a standard cell supplied by any maker will be accompanied by a certificate giving the variation of the E.M.F. per degree centigrade.

There are, of course, no means taken for the prevention of polarisation.   The cell is not intended to generate currents but to be a standard of E.M.F.   Thus in practice one should always use a high resistance—something not less than 10,000 ohms—in series with the cell, to ensure that in no case can a current pass through the cell of sufficient strength to produce polarisation.   And again, since the cell does not generate appreciable currents and is hermetically sealed, it should be practically everlasting.

The Cadmium Cell.   Another form of standard cell, and one likely to be adopted officially, is that known as the cadmium cell designed by Mr Weston.   In this cell cadmium is used in place of the zinc of the Clark cell, and its chief advantage lies in the fact that the variation of the E.M.F. with the temperature of the cell is very small compared with the other standard.

Fig. 17 illustrates a form of cadmium cell, made up in

the H form suggested by Lord Rayleigh. The diagram is self-explanatory.

The E.M F. of the cadmium cell is 1·019 volts at 20° C. and its temperature variation of E.M.F. is only one-tenth of that of the Clark cell.

The ordinary cells used in everyday testing work are used for the purposes of generating small currents and not as standards of E.M.F. The majority of these polarise very readily and, as the student is aware, this reduces their

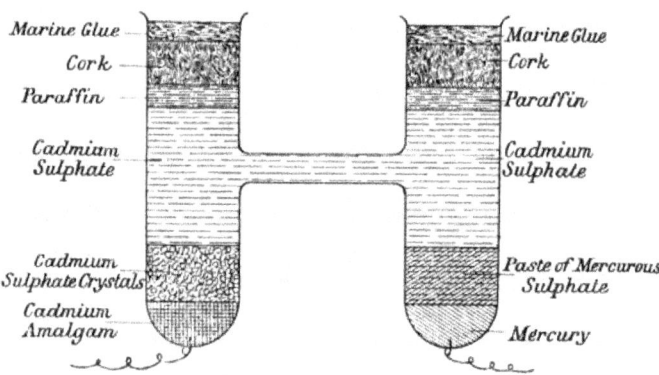

Fig. 17.

effective E.M.F. by setting up a back E.M.F. ; and increases the internal resistance of the cell. The Daniell cell is the most constant under working conditions and has an E.M.F. varying from 1·07 to 1·114 volts, depending upon the density of the copper sulphate and the zinc sulphate solutions. The E.M.F.s of the various forms are : Grove's 1·95 volts, Bichromate 2·1 volts, Leclanché 1·46 volts, Bunsen 1·90 volts.

A large battery of cells is useful for testing purposes. A battery of 100 Leclanché cells in series for example would give an E.M.F. of 146 volts. Such a battery would be useless for commercial purposes such as lighting or motor driving, for the internal resistance being very high the efficiency would be extremely low. And the cost per unit would be some

four or five times greater than that supplied by the most
expensive electric power supply company.

Drop in Potential down a wire. The difference of
potential between two points in a circuit is measured in
practice by an instrument called a *voltmeter*. The underlying
principles of the measurement generally cause the student no
little difficulty, with the result that amperes are only too
often confused with volts, and voltmeters are treated like
ammeters.

When a current is flowing along a wire there can not
be two points on that wire which have the same potential.
Further, as the resistance between two points increases, then
the difference of potential between them must also increase
if the current is to remain the same. This follows from Ohm's
law, and this is the first point to be mastered. Hence if AB
in Fig. 18 be a long wire whose resistance is proportional to

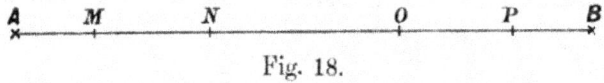

Fig. 18.

its length, and through which a constant current is flowing,
then there will be a difference of potential between A and
B which is equal to the product of the current in the wire
and the resistance between A and B.

Now if the current is flowing from A to B it follows that
the potential of A must be higher than that of B, and, indeed,
must be higher than that of any other point such as M, N, O,
or P, between A and B. And again all the points must
be at a higher potential than B. Further the point M must
be at a higher potential than the point N, N must be higher
than O, and so on.

On this diagram the length AM is equal to the length PB;
and MN is equal to OP. Since the wire is assumed to be a
uniform one, lengths will be proportional to resistances, and
thus the resistance between the points AM is equal to that

between P and B; and the currents in these are equal since they are in series.

Now to maintain equal currents in equal resistances, equal E.M.F.s are required—again following from Ohm's law. Therefore the *difference* of potential between AM must be equal to that between PB.

It must be appreciated thoroughly however that this does not imply that the *actual potentials* at the respective points are equal. They are not, nor can they be if it is conceded that there cannot be flow of electricity unless there is a difference of potential.

It may be supposed, for example, that the difference of potential between A and B is represented by 10 volts. Now the point B may be connected to earth—so that its potential would be zero. The actual potential of A would then be 10 volts. Let it be further supposed that the resistance of the wire AB is 20 ohms; therefore the current passing through it is $0{\cdot}5$ ampere. Let AM and PB each have a resistance of 4 ohms; and MN and OP a resistance of 2 ohms.

Therefore the difference of potential between

$$A \text{ and } M \text{ must} = 4 \times 0{\cdot}5 = 2 \text{ volts.}$$
$$M \text{ and } N \quad ,, \quad = 2 \times 0{\cdot}5 = 1 \text{ volt.}$$
$$O \text{ and } P \quad ,, \quad = 2 \times 0{\cdot}5 = 1 \text{ volt.}$$
$$P \text{ and } B \quad ,, \quad = 4 \times 0{\cdot}5 = 2 \text{ volts.}$$

Hence it must follow that the actual potential of the point M must in this case be 8 volts, and N must be 7 volts, O must be 3 volts and P must be 2 volts.

Between N and O therefore the resistance must be the difference of potential divided by the current. That is $\dfrac{7-3}{0{\cdot}5} = 8$ ohms. And of course this agrees with the assumed resistances of the whole wire and the remaining parts.

Actual potential does not come into electrical measurements. It is *difference of potential* which is required, and the above illustrations have only been introduced to emphasise this. There is the same E.M.F. between P and B as there is between A and M. And the fact is established that as the resistance between two points increases, so does the potential difference increase *if a constant current be flowing.* This may be stated in another way, that as the resistance between two points increases so *must* the difference of potential between them increase *in order to keep a constant current flowing.*

Electrostatic Voltmeter. The ideal mode of measuring E.M.F.s is that based on statical effects. If two insulated conductors be connected to two points at different potentials, they will become equal in potential to the points. Now in order to bring about this some electricity must pass to them or from them as the case may be, but when once the potentials are reached there will be no further transference of electricity provided that a constant electrical state is maintained. Moreover the quantities necessary to adjust the potentials of the insulated conductors would be very small indeed compared with the unit of quantity called the coulomb—so small indeed as to be negligible.

Now the two insulated conductors being at different potentials, there will be a tendency between them for equalisation, and if this tendency can be measured in any way, it will give a measure of the difference of potential between the conductors.

This is the principle employed in the type of voltmeters known as electro-static, which are practical forms of the *quadrant electrometer.* Fig. 19 is an illustration of a Kelvin electro-static voltmeter designed for measuring high E.M.F.S.

In the electro-static voltmeter two brass butterfly-shaped plates A and B, Fig. 19, are placed parallel to each other. They are connected together, and a butterfly needle N of aluminium is pivoted at PP' so that it can swing in a plane parallel to A and B, and in between them. The needle is insulated from the plates or "quadrants" as they are often called and is connected, *via* its pivots, to one terminal of the instrument, the quadrants being connected to the other. Generally the needle is connected to the " case," which is of iron, and the quadrants are specially insulated from it.

To the upper end of the needle a pointer is fixed which moves over a scale as shewn. At the bottom of the needle there are two small projecting arms. These are screwed, and two small nuts K and K' can be moved in either direction

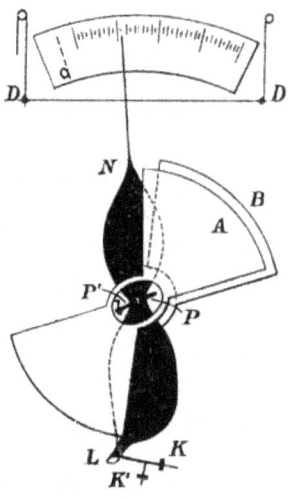

Fig. 19.

along the threads. These serve as counterweights for adjustment of the zero position of the needle. There is also a small loop L on which may be hung different weights (supplied with the instrument), by means of which the controlling forces (gravity control) may be varied, and consequently the degree of sensitiveness of the instrument varied.

When the terminals are connected to two points at different potentials the needle tends to move in between the quadrants, the greater the difference of potential the greater will be the tendency. If however a different weight be hung on the loop L the actual movement will be greater or less for the same potential differences as the weight is less or greater.

The scale is marked off in such a way that with one weight, say W_1, the potential difference will be *scale divisions* × 50 *volts*. With W_2 it will be *scale divisions* × 100 *volts*, and so on. The special value of this lies in the fact that an "open scale" may be used over a considerable range of potential difference.

The quadrants are connected to one and the needle is connected to the other of the two points between which the difference of potential is to be measured. The needle and quadrant tend to equalise potential, and this tendency causes the needle to be attracted into the quadrants with a force proportional in some way to the difference of potential between them. Thus it only remains to mark off the scale, over which the pointer moves, in volts by direct comparison with some standard—and this is done in practice by comparing the voltmeter with a quadrant electrometer which in turn can be standardised with a Clark cell.

These electro-static voltmeters can be used for continuous or alternating E.M.F. measurements. They do not absorb any electrical energy, and consequently do not alter the difference of potential of any two points by being connected to them. They are not affected by magnetic fields ; they are not heated by the charges on quadrants and needles ; and appear to approach as nearly to the ideal conditions as could be hoped for in a practical instrument.

The instrument described and illustrated above is only suitable for high E.M.F.s ; but by increasing the number of quadrants and needles the sensitiveness may be increased to almost any degree. When this is done the needles are generally suspended so that they can move in a horizontal plane about a vertical axis, the suspension being a fine silver wire. The needles are all parallel to one another with a common axis. The quadrants—also horizontal—are parallel, and each needle can move between a pair of quadrants. The instrument is called a multi-cellular voltmeter, and is illustrated diagrammatically by Fig. 20. The pointer moves over a horizontal circular scale at the top of the instrument or is bent at the end to move over a vertical scale. The needle is rendered dead-beat by means of a pierced ivory disc at the bottom of the needle, which is immersed in a vessel of oil

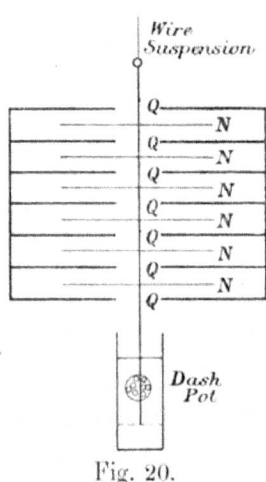

Fig. 20.

at the base. This arrangement is called a dash pot. The High Tension voltmeter is not dead-beat, but an insulated rod is arranged at the top—marked *D* in Fig. 19—by means of which the movement of the needle can be arrested.

ELECTRO-MAGNETIC VOLTMETERS.

The electro-static voltmeters have an infinite resistance, and a current does not pass through them. However there are also voltmeters which are used to measure E.M.F.s by measuring small currents—electro-magnetic voltmeters—and these are frequently sources of mystery to beginners in electrical measurements.

Conditions of a good voltmeter. Now a voltmeter should fulfil certain conditions. It should be direct reading ;

dead-beat ; sufficiently sensitive for the degree of accuracy aimed at ; and of such a *high resistance* that its connexion across two points of a circuit does not affect appreciably the resistance between those points. This last condition is of supreme importance, and it is in this particular that a volt-meter differs totally from an ammeter.

When an electro-magnetic voltmeter is connected across two points in a circuit between which there is a difference of potential, a current will pass through the coil of the instrument. This current will be proportional to the differ-ence of potential between the points, and will produce a magnetic effect by means of which its strength can be measured.

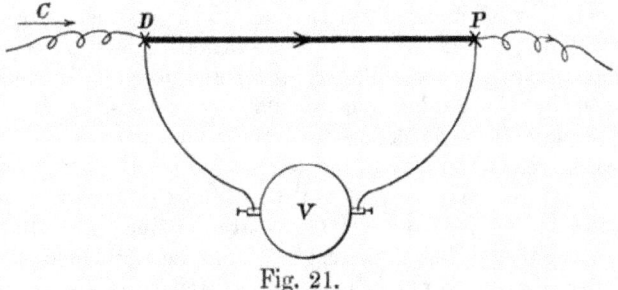

Fig. 21.

Let D and P on Fig. 21 be two points in a circuit in which a current is flowing. It is desired to measure the difference of potential between these points by means of a voltmeter. Now the difference of potential between D and P will depend upon the current and the resistance between them. If either be altered the difference of potential will be altered.

An electro-magnetic voltmeter V is connected to D and P. Now a portion of the current C will be shunted through the voltmeter at D—or it may be said that the current will divide at D, part going through V and the rest along the resistance DP. The strengths of these parts according to Ohm's law will be inversely proportional to the resistances of the two paths. Hence if the resistance of V be very great only a very small portion of the original current will be "shunted" through it. Again, if V be very great the combined resistance of the voltmeter and the resistance between DP will not be appreciably *less* than the resistance between DP without the voltmeter.

Thus the difference of potential between D and P will remain practically unaltered by the connexion of the voltmeter since the current strength and the resistance are unaltered. But it can be seen that if V had a low resistance—comparable to the resistance of DP—the total resistance of the circuit would be *lessened* by connecting it up to D and P, the current strength in DP would be altered and the difference of potential between D and P would be different from that which it was intended to measure. This matter is treated again on page 72 after the consideration of joint resistance when it can be more fully discussed and more easily understood by the reader, who should nevertheless be able to appreciate in some degree from the above the necessity for having voltmeters of high resistance.

Now the small current which is shunted through a voltmeter has to pass through many turns of fine wire. It will be remembered that the strength of the magnetic field at the centre of a coil is proportional to the strength of the current and to the number of turns of wire through which it passes. Hence in a voltmeter, although it is of high resistance and although the strength of the current which passes through it is small, yet the magnetic field in its coil is comparable to the field of an ammeter coil because of the very large number of turns of wire through which the current passes.

Thus in an ammeter coil the field strength is produced by a large current through few turns; and in a voltmeter by a relatively small current through many turns.

Mechanically there need be no difference in construction of ammeters and voltmeters except the difference of the coils. The voltmeter must have many turns of wire and this wire can be of small gauge since it will only have to carry a very small current.

4—2

There are two types of electro-magnetic voltmeters—and these correspond exactly to the " A " and " B " types of electro-magnetic ammeters. There is therefore no need to enter into details of their construction—the reader can substitute mentally a fine wire coil of many turns for the thick wire coil of the ammeter, and a scale graduated in volts for the ampere scale. It can be seen that the small current in the voltmeter will depend entirely upon the difference of potential at its terminals, and that the scale can be graduated therefore in volts.

Coil type Voltmeter. The " B " type of electro-magnetic voltmeter and ammeter consists of a suspended coil and a fixed magnet, after the manner of a D'Arsonval galvanometer.

The movable coil voltmeter consists of a coil of fine copper wire having a high resistance, wound on a rectangular frame or " former " of aluminium. The frame is pivoted and mounted on jewels at each end of its long axis. Sometimes the axis is arranged vertically and in other forms it is arranged horizontally. At each end of the pivot the inner end of a hair spring (made of phosphor-bronze usually) is attached, whilst the outer end is fixed. The end of each spring is insulated from the pivot and connected on to one end of the coil, and it is by means of these springs that the current is passed to and from the coil. At the same time the springs supply the necessary controlling forces.

A strong permanent magnetic field is supplied by one or two steel horseshoe magnets, so placed that when the coil is controlled only by the springs and is at rest its plane will be parallel to the magnetic field between the poles, and in the line across the centre of the field.

The field between the poles is strengthened by means of a cylinder of soft iron which is *fixed* inside the coil but quite free from it, with as small an air gap as is consistent with free turning of the coil. Further, the air gaps between the magnet poles and the coil are as small as possible and the pole-pieces are curved with the object of securing as strong and uniform a magnetic field as is possible with a given magnet.

An aluminium pointer is fixed to the axis of rotation of the coil, and this moves over a scale which is graduated in volts.

The coil being of very high resistance there will only be a small current set up in it when it is connected to the two points whose difference of potential is to be measured. This will not be sufficient to cause any appreciable heating of the hair springs or of the coil itself.

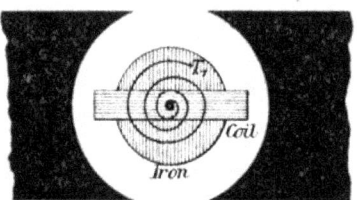

Fig. 22.

The magnetic effect of this current, which will be proportional to the difference of potential being measured, will tend to turn the coil through 90°. The scale is marked by comparison with a standard voltmeter. Fig. 22 illustrates the coil type of voltmeter. The figure is a diagrammatic plan.

To make the voltmeter of sufficiently high resistance a fixed non-inductive resistance is placed inside the instrument and is connected in series with the movable coil.

Coil type Ammeter. It will be seen readily that it would be impossible to make a movable coil of thick wire capable of carrying a big current—not because of the coil itself, but because of the controlling springs and connecting wires. These must be fine to allow for the delicate movements of the coil and thus a direct form of movable coil ammeter cannot be made. In practice a movable coil ammeter is really a *shunted voltmeter*, being the same in construction, having a coil of fine wire and fine controlling springs and connexions. But the ends of the coil are connected to a low resistance — shunted across it as shewn in Fig. 23. T_1 and T_2 are the terminals of the "ammeter." The ammeter is shewn connected up in a circuit. The low resistance does not affect the total resistance of the circuit. At the same time the current which passes through it will cause a difference of potential between T_1 and T_2 which will be proportional to the strength of that current. Hence the fine wire coil will have a current passing through it which is proportional to the difference of

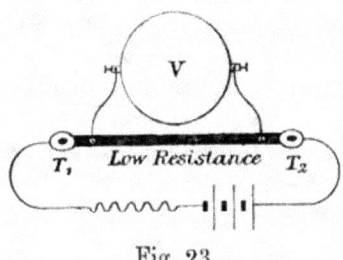

Fig. 23.

potential between T_1 and T_2, which in turn is proportional to the strength of the current in the low resistance between T_1 and T_2. Hence the scale under the pointer of the coil can be graduated in amperes.

This may be illustrated very simply in this way. If a voltmeter be shunted across a resistance of one ohm through which currents of different strengths could be passed, the voltmeter would indicate the difference of potential across the ends of the ohm resistance.

That is to say E volts $= C$ amps $\times R$ ohm.

But R in this case $= 1$ ohm.

$\therefore$ The reading in volts is numerically equal to the current strength in amperes of the currents passing through the 1 ohm coil.

Similarly it can be seen that if the voltmeter be shunted across a resistance of ·01 ohm, the reading of the voltmeter would be equal to one-hundredth of the number of amperes passing through the resistance, and so on.   This is the principle of the movable coil ammeter.

The movable coil instruments have many advantages over iron needle types.   They are practically unaffected by stray magnetic fields and are very accurate   They cannot be used except for one direction of the current and are therefore useless for alternating currents.   But their reliability and constancy render them preferable at all times for continuous current measurements.

In addition to electro-magnetic and electro-static voltmeters, there are also hot wire voltmeters working on identical principles to the hot wire ammeters.   These all depend for their working upon the fact that a current produces a heating effect, and that a wire expands on being heated.   Hot wire voltmeters can be used for alternating or continuous E.M.F.s, and are unaffected by magnetic fields.   The wire is liable to physical deterioration.

CHAPTER IV.

RESISTANCE AND ITS MEASUREMENT.

EVERY electrical circuit offers resistance to the transmission of an electric current. That is merely another way of saying that nothing is known which will transmit electricity perfectly—that some energy is always necessary to urge the current through the medium desired. The amount of such energy will depend upon the amount of resistance offered and upon the strength of the current it is desired to maintain, and this is fully discussed in the next chapter.

The nature of the resistance offered to a current of electricity is unknown, and it is dangerous to use analogies for illustration, for the very good reason that there are none which hold good throughout. The resistance offered to the flow of water in a pipe is an oft-used analogy, and is useful perhaps in the illustration of certain points ; but the fact remains that electricity cannot be compared to water without leading up to serious fallacies ultimately.

In the light of what is known it can only be said that the strength of a current of electricity is determined by the E.M.F. acting and by another factor called the resistance. This is Ohm's law, and the unit of resistance must be that resistance which is offered when a current of unit strength is maintained through it by a unit of E.M.F.

The Ohm. The Board of Trade definition of the practical unit of resistance is "the resistance of a uniform column of mercury 106·3 cms. long and one square millimetre in area of cross section at a temperature of 0° C." This unit is called an *ohm*.

Now although this is called a practical definition, it is nevertheless extremely impractical, for the reader will see the difficulty of making an ohm resistance from the above definition. The adjustment of such a column is almost impossible, and it should be stated at once that the definition is not fundamental but has been derived backwards from the absolute definition of the unit of resistance. The practical standards of resistance do not take the form of mercury columns except as ultimate reference standards, but are made, as will be seen later, of wire.

Laws of Resistance. Matter may be divided into two classes so far as the transmission of electricity is concerned ; those substances which conduct electricity well, and those which conduct badly. Of course the classification is purely relative, and the terms *conductors* and *insulators* are given to the two classes respectively. There are no perfect conductors and no perfect insulators, although there are substances which offer such great resistance that, at present, it cannot be measured.

Electrical conductors are generally made in the form of a wire of circular cross section. This is chiefly a matter of convenience in manufacture and a consequent keeping down of cost. They may be made of any cross sectional shape, for the shape of the cross section does not affect the conductivity of the substance.

The resistance of a conductor of a given substance depends upon the *length* of the conductor and upon the *area* of its *cross section*. It is directly proportional to the length and inversely proportional to the area of cross section. If the conductor has a variable cross sectional area, then its resistance varies inversely as the *mean area of cross section*. These facts admit of simple experimental proof, the details of which may be readily thought out by the student, who should endeavour to put his ideas to the test of experiment.

The resistance of a conductor also depends upon the nature of the material used. If a series of wires be made of equal length and equal area of cross section but of different substances they will be found to have different resistances. This is expressed by saying that different substances have different *specific resistances*, and the *specific resistance of a substance* is defined as *the resistance of a unit length of that substance, having a unit area of cross section.*

The units of length and area generally used are the centimetre and square centimere on the metric system, and the inch and square inch on the British system. Below is given a table shewing the specific resistance of some of the most important substances in terms of both sets of units. The resistances are given in ohms.

The values of these resistances are often given in *microhms*, that is in millionths of an ohm. They are often quoted also as being so many ohms or microhms per cubic inch or per cubic centimetre. The student is warned against these statements because they are apt to imply that the resistance of a conductor varies as its volume. It is a common error, and should be guarded against. The resistance of a cubic inch of copper is quite unknown, for a cubic inch of copper might be in the form of a wire a mile long and $\frac{1}{63,360}$ sq. inch in area of cross section; or it might be a conductor half an inch long and 2 sq. ins. in area of cross section. The resistances would be quite different.

The statement that the specific resistance of copper is ·000,000,66 ohm per cubic inch is intended to convey to the mind the fact that with a cube of copper of 1 inch side, the resistance between two parallel faces is ·000,000,66 ohm. It must be admitted that the definition is not precise, but it is generally used and the reader will be able to interpret it correctly. In the examples given in this volume specific resistances will be quoted as so many ohms per inch per sq. inch, or ohms per cm. per sq. cm., according to the units in use.

Referring to the table of specific resistances it is seen that silver is the best conductor, closely followed by copper. The latter substance is generally used for conductors, because a

Table of Specific Resistances.

Substance	Resistance of 1 cm. length having area of cross section of 1 sq. cm.	Resistance of 1 in. length having area of cross section of 1 sq. in.
Metals		
Silver annealed ...	0·000001468 ohms	0·000000578
Copper annealed ...	·000001561 „	·000000615
Copper (cables) ...	·000001676 „	·00000066
Gold	·000002197 „	·000000865
Aluminium	·000002665 „	·000001049
Iron	·000009065 „	·000003569
Platinum	·000010917 „	·000004298
Mercury	·00009407 „	·00003703
German Silver ...	·000030 „	·00001181
Platinoid	·0000417 „	·00001641
Manganin	·000042 „	·00001653
Liquids		
Water at 4° C. ...	8,978,000 „	3534648
„ 11° C. ...	335,000 „	131889
Sulphuric acid		
5% acid at 15° C. ...	4·81 „	1·89
Copper sulphate		
(saturated) at 10° C.	28·9 „	11·37
Non-Metals		
Carbon (arc lamp) ...	·003 to ·004 „	0·00118 to 0·00157
Insulators		
Shellac	9000×10^6 „	3543×10^6
Paraffin	$34,000 \times 10^6$ „	13385×10^6
Ebonite	$28,000 \times 10^6$ „	11023×10^6
Gutta Percha at 24° C.	450×10^6 „	177×10^6
Mica	83×10^6 „	33×10^6
Glass (soda-lime) 20° C.	$8·96 \times 10^6$ „	$3·5 \times 10^6$
Glass (Crystal) below 40° C.	∞	∞

given length of a given resistance can be produced at a lower cost than with any other substance.

Alloys are seen to have relatively high specific resistances.

The specific resistance of an alloy is usually much higher than that of any of the pure metals of which it is composed. *Platinoid* is an alloy of copper, zinc, nickel and tungsten. *Manganin* is an alloy of copper, nickel and manganese. Alloys have small temperature variations and are invariably used in the manufacture of resistance standards, resistance boxes, and the like.

Resistance Laws continued.   Since the resistance of a conductor varies directly as its length, and as its specific resistance, and inversely as its area of cross section, it follows therefore that

$$R \propto \frac{l \times s}{a},$$

when R = resistance of a length l having an area of cross section a and a specific resistance s.   If R be in *ohms*, l in *centimetres*, a in *square cms.* and s in *ohms per cm. per sq. cm.*, it follows that

$$R = \frac{l \times s}{a}.$$

The same expression will hold good if l be in *inches*, a in *square inches*, and s in *ohms per inch per square inch*.

One can always convert units in one system to equivalents in another system.   For example, l might be in inches and a in sq. inches, whilst s is given in ohms per cm. per sq. cm. If R was required, it would be necessary firstly to change l and a into centimetres and sq. cms. respectively; or to change s to ohms per inch per sq. inch.

Now 2·54 centimetres are equivalent to 1 inch length. Therefore 2·54 × 2·54 sq. centimetres will be equivalent to 1 square inch area.   That is 6·4516 sq. cms. = 1 square inch.

Hence l inches × 2·54 = L centimetres,

and a sq. inches × 6·45 = A sq. cms.

Similarly since 1 inch = 2·54 cms., therefore

$$1 \text{ cm.} = \frac{1}{2·54} \text{ inch} = 0·3937 \text{ inch,}$$

and

1 sq. cm. $= 0.3937 \times 0.3937$ sq. inches $= 0.15499969$ sq. inch,

which may be taken as **0.155 sq. inch.**

Thus L cms. $\times 0.3937 = l$ inches,

and A sq. cms. $\times 0.155 = a$ sq. inches.

If the resistance of a given length of a substance having a known area of cross section be determined, the specific resistance may be readily calculated. For since

$$R = \frac{l \times s}{a}, \quad \therefore \ s = \frac{a \times R}{l}.$$

For example, the specific resistance of copper is $0.000,000,66$ ohm per inch per sq. inch. What is its specific resistance per cm. per sq. cm. ?

Now in this case $R = 0.000,000,66$,

$$l = 1 \text{ inch,}$$

$$a = 1 \text{ sq. inch.}$$

It is required to find the specific resistance in ohms per cm. per sq. cm.

To do this l must be expressed in cms. and a in sq. cms.

$\therefore l = 1 \times 2.54 = 2.54$ cms., and $a = 6.45 \times 1 = 6.45$ sq. cms.

Thus $s = \dfrac{6.45 \times .00000066}{2.54}$

$= 0.000,001,676$ ohms per cm. per sq. cm.

Another example may be taken as follows : A wire of circular cross section is 210 cms. long and has a diameter of 0.03 cm. Its resistance is found to be 12.47 ohms. What is its specific resistance (i) in ohms per cm. per sq. cm., (ii) in ohms per inch per sq. inch?

(i) $R = 12.47$ ohms,

$$l = 210 \text{ cms.}$$

The area of a circle $= \pi \times$ radius^2,

$\therefore a = \pi \times \cdot015 \times \cdot015$ sq. cms. $= \cdot000707$ sq. cm.

$$\therefore s = \frac{a \times R}{l} = \frac{\cdot000707 \times 12\cdot47}{210} = \cdot000042 \text{ ohm}$$

$= 42$ microhms per cm. per sq. cm.

(ii) $K = 12\cdot47$ ohms,

$l = 210$ cms. $= 210 \times \cdot3937$ inches,

$a = \cdot000707$ sq. cms. $= \cdot000707 \times \cdot155$ sq. inches.

$\therefore s$ in ohms per inch per sq. inch $= \dfrac{a \times R}{l}$

$$= \frac{\cdot000707 \times 0\cdot155 \times 12\cdot47}{210 \times \cdot3937}$$

$= \cdot0000165$ ohm

$= 16\cdot5$ microhms per inch per sq. inch.

If the specific resistance in terms of one set of units be known, it can be more simply determined in terms of the other set, as follows :

$$R = \frac{l}{a} \times s.$$

Let $l = 1$ cm. and $a = 1$ sq. cm.

$\therefore l = \cdot3937$ ins. and $a = \cdot3937 \times \cdot3937$ sq. inches.

$$\therefore R = \frac{\cdot3937}{\cdot3937 \times \cdot3937} \times s = \frac{1}{\cdot3937} s = 2\cdot54 s,$$

where $s =$ specific resistance in ohms per inch. per sq. inch.

If $l = 1$ inch and $a = 1$ sq. inch,

$\therefore l = 2\cdot54$ cms. and $a = 2\cdot54 \times 2\cdot54$ sq. cms.

$$\therefore R = \frac{2\cdot54}{2\cdot54 \times 2\cdot54} \times s = \frac{1}{2\cdot54} s = \cdot3937 s,$$

where $s =$ ohms per cm. per sq. cm.

In short the specific resistance s in ohms per cm. per sq. cm. is the same as $0\cdot3937s$ in ohms per inch per sq. inch ; or if s be in ohms per inch per sq. inch, then $2\cdot54s$ is the specific resistance in ohms per cm. per sq. cm.

Variation of Resistance with Temperature. The resistance of a substance also varies in some way with its temperature. The alteration of resistance with temperature is one of the most important effects of heat. It is found that the resistance of the pure metals increases uniformly with the temperature. *The alteration in resistance of a unit resistance at 0° C. for a rise of temperature of 1° C. is called the temperature coefficient of resistance.*

Thus a length of copper wire which has a resistance of 1 ohm at 0° C. is found to have a resistance of 1·00428 ohms at 1° C. The increase, viz. 0·00428 ohm, is called the temperature coefficient of resistance of copper.

The resistance of any coil of a metallic conductor at any temperature $T°$ C. will be given by the expression

$$R_T = R_0\,(1 + aT),$$

where R_T is the resistance at the temperature $T°$ C., and R_0 is the resistance at 0° C. and a is the temperature coefficient of resistance of the material of the conductor.

In the following table the temperature coefficients of various substances are given.

Table of Temperature Coefficients.

Substance	Temperature coefficient
Silver	·004
Copper	·00428
Gold	·00377
Aluminium	·00435
Iron	·00625
Platinum	·003669
Mercury	·00072
German silver	·002
Platinoid	·00022
Manganin (0—10° C.) ...	·000025
Manganin (10—30° C.) ...	Practically zero
Non-metals and electrolytes	Negative temperature coefficients
Carbon (arc lamp) ...	− 0·00052

The temperature coefficient of alloys is always less than that of any of their component pure metals and some alloys are found to have practically no alteration of resistance with temperature. This fact, combined with their high specific resistance, renders them especially suited for the manufacture of resistance standards.

The variation of resistance of a metal with temperature can be very simply shewn as follows. A piece of iron wire— about number 20 s.w.g.—of 2 metres length is wound into a

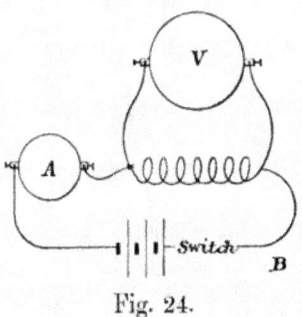

Fig. 24.

spiral as shewn on Fig. 24. This is joined in series with an ammeter and a battery ; and a voltmeter is connected across its ends.

If the iron be heated up with a Bunsen flame the reading of the voltmeter will increase even though the reading of the ammeter decreases. The latter fact shews that the current strength has diminished, due to increased resistance, whilst the increased reading of the voltmeter indicates that the resistance has increased so much that with even a smaller current a greater difference of potential must be acting at its ends. Of course in each case the resistance

$$= \frac{\text{difference of potential}}{\text{current strength}}.$$

It is seen that the average temperature coefficient of most of the pure metals is 0·000366 for all ordinary ranges of

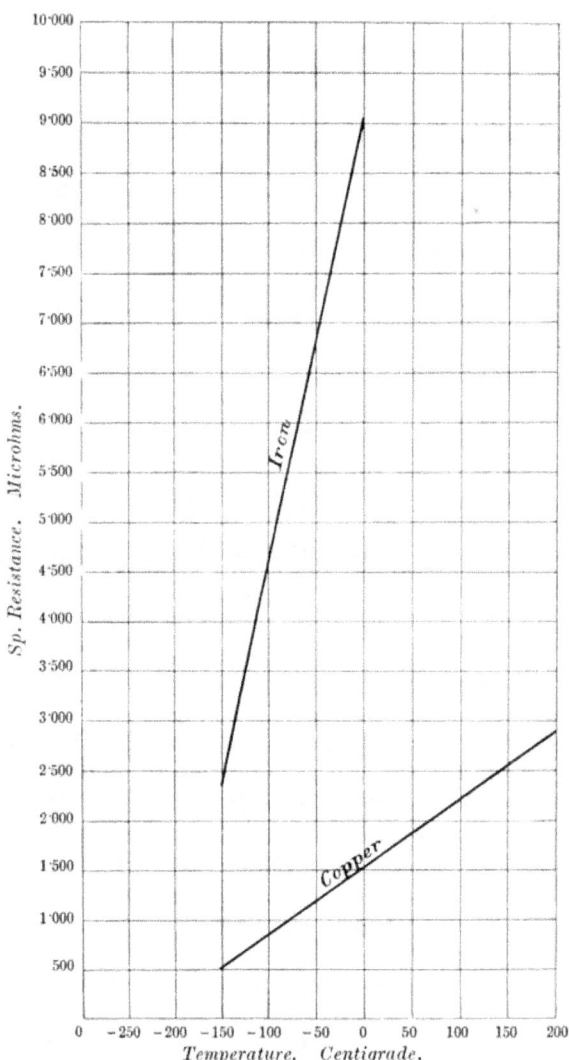

Fig. 25.

temperature. The value of this coefficient is the same as the coefficient of expansion of air, and as a consequence it would appear that if pure metals were cooled down to a temperature of $-273°$ C. their resistance would be approximately zero. ($\cdot000366 = \frac{1}{273}$.)

Fig. 25 is a graphical illustration of the relationship between the specific resistance and temperature of two substances.

The resistance of non-metals is found to decrease as the temperature increases. They are said to have a negative temperature coefficient of resistance and this is indicated by prefixing the minus sign to the value of the coefficient. Insulators fall under the heading of non-metallic substances, and since their resistances decrease with an increase in temperature it is seen that considerable trouble may arise due to the insulating material of a cable becoming hot. For example, the specific resistance of crystal glass below $40°$ C. is infinity; but at $105°$ C. it is reduced to 11 million ohms.

The resistance of some insulators at a high temperature is so low that it would be dangerous to use them as insulators of cables for high E.M.F. power distribution.

Liquids which are decomposed by the passage of an electric current are called electrolytes, and these are also found to have a negative temperature coefficient, even though they be solutions of metallic salts. It follows therefore that the internal resistance of a voltaic cell decreases as its temperature increases. It may here be noted that the E.M.F. of a cell is altered by alteration of temperature, but of course the variations of E.M.F. and internal resistance are in no way related to one another.

Oils, generally, are good insulators; but they have also a negative temperature coefficient of resistance.

The specific resistances of gases are too great for measurement, and they are consequently put down as infinity. The fact that a spark may be made to jump across an air gap with a finite E.M.F. must not be regarded as a necessary refutation of this, for there is evidence that a high potential difference between two points separated by air or other dielectric produces a physical strain in the dielectric. This strain ultimately breaks down the dielectric resistance and a spark passes; but this transference of electricity cannot be said to follow Ohm's law.

Joint Resistance of Resistances in Series. Let AB, BC and CD in Fig. 26 be three resistances which are connected in series. Let the resistance of $AB = r_1$; of $BC = r_2$;

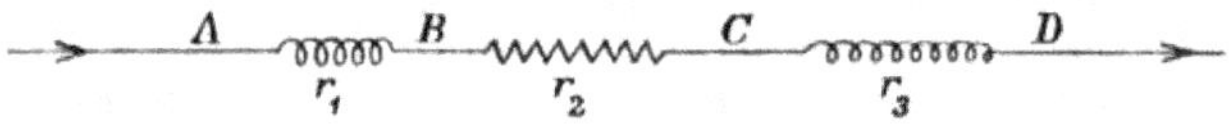

Fig. 26.

of $CD = r_3$; and the whole resistance between A and $D = R$. Let the current in these $= C$ and the total difference of potential between A and $D = E$.

It follows that E must equal the sum of the differences of potential across r_1, r_2, and r_3.

$$\therefore\ E = Cr_1 + Cr_2 + Cr_3 = C(r_1 + r_2 + r_3).$$

Now the total resistance $R = \dfrac{\text{total E.M.F. } E}{\text{total current } C}$,

$$\therefore\ R = \frac{C(r_1 + r_2 + r_3)}{C} = r_1 + r_2 + r_3.$$

The combined resistance of any number of resistances joined up in series is equal to the sum of those resistances.

Joint Resistance of Resistances in Parallel. Now let the resistances be connected in parallel, as shewn in Fig. 27. Since they are all connected at one end at the point A, and at the other end at the point B, it follows that the difference of potential across each of the resistances is the same; and is equal to the total difference of potential across AB. Let this be denoted by E.

It is clear therefore that unless the resistances r_1, r_2 and r_3

5—2

be equal the currents in these resistances will not be equal. Let these be denoted by c_1, c_2 and c_3 respectively.

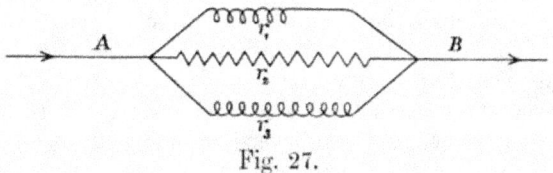

Fig. 27.

$$\therefore \ c_1 = \frac{E}{r_1} \ ; \ c_2 = \frac{E}{r_2} \ ; \ \text{and} \ c_3 = \frac{E}{r_3}.$$

Let the current at $A = C$. The current at B will also be equal to this.

$$\therefore \ C = c_1 + c_2 + c_3.$$

$$\therefore \ C = \frac{E}{r_1} + \frac{E}{r_2} + \frac{E}{r_3} = E\left(\frac{1}{r_1} + \frac{1}{r_2} + \frac{1}{r_3}\right).$$

The total resistance between $AB = R = \dfrac{E}{C}$.

$$\therefore \ R = \frac{E}{E\left(\dfrac{1}{r_1} + \dfrac{1}{r_2} + \dfrac{1}{r_3}\right)} = \frac{1}{\dfrac{1}{r_1} + \dfrac{1}{r_2} + \dfrac{1}{r_3}}.$$

This is sometimes expressed as

$$\frac{1}{R} = \frac{1}{r_1} + \frac{1}{r_2} + \frac{1}{r_3}.$$

Now the *conductivity* of a substance is the inverse of its *resistivity*, as specific resistance is sometimes called, and it may be stated therefore that the total conductance of an arrangement of any number of resistances in parallel is equal to the sum of their individual conductances.

It follows, too, that the total resistance of a system of parallel resistances must be always less than the least of the individual resistances of that system. This is a point over which students often slip, but it should be easy to grasp. Every additional resistance connected in parallel must increase the equivalent cross sectional area of the system—therefore the total resistance must be lowered. The combined resistance of 1 ohm and 1,000,000 ohms in parallel must be less than 1 ohm, although the alteration would be so small as to be negligible.

Again, since there must be equal drops of potential through the separate resistances of a parallel system, the current in each component must be in inverse ratio to its resistance. In the case just quoted the ratio of the currents in the 1 ohm and 1,000,000 ohm components would be 1,000,000 : 1.

Effect of Voltmeters in Shunt. It will be advisable here to refer back to the use of voltmeters and the necessity for their high resistance. Let Fig. 28 represent a circuit

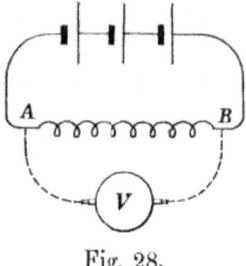

Fig. 28.

consisting of a battery whose E.M.F. is 4 volts and whose resistance is 1 ohm ; and a resistance AB of 3 ohms. From Ohm's law the current in the circuit will be equal to 1 ampere.

From the same law it follows therefore that the difference of potential across AB must be $3 \times 1 = 3$ volts.

This should be registered by the ideal voltmeter. But if a voltmeter whose resistance is, say, 3000 ohms be introduced and connected up to AB there would be some alteration produced.

The total resistance of the circuit would now be the sum of the battery resistance and the combined resistance of voltmeter and AB. The combined resistance of AB and the voltmeter would be

$$\frac{1}{\dfrac{1}{3}+\dfrac{1}{3000}} = \frac{1}{\dfrac{1000+3}{3000}} = \frac{3000}{1003} = 2\cdot991 \text{ ohms.}$$

$\therefore$ The correct total current $C = \dfrac{E}{\text{Total } R} = \dfrac{4}{1 + 2\cdot991}$

$$= \frac{4}{3\cdot991} = 1\cdot00225 \text{ amperes,}$$

the difference being 2 parts in one thousand or $0\cdot2$ °/$_\circ$.

The total current being $1\cdot00225$ amps. and the resistance between A and B being $2\cdot991$ ohms, it follows therefore that the difference of potential between A and B will be

$$2\cdot991 \times 1\cdot00225 \text{ volts}$$

$$= 2\cdot9977 \text{ volts.}$$

Hence without the voltmeter the true difference of potential $= 3$ volts.

And with the voltmeter the true difference of potential $= 2\cdot9977$ volts.

The error produced by the voltmeter resistance is therefore rather less than 1 part in 1000 or less than $0\cdot1$ °/$_\circ$. This is sufficiently accurate for most practical purposes.

It can be seen, and the student should work out some cases for himself, that a voltmeter of smaller resistance would produce a greater error—an error which would soon become too great for most ordinary measurements. And on the other hand, as the voltmeter resistance is increased the error becomes less and less, until finally the ideal conditions are realised when the voltmeter resistance is infinity.

Appreciation can now be given to this condition of a voltmeter, that it shall have such a *high resistance* that its connexion to two points of a circuit *does not alter* to any practical extent *the total resistance* of that circuit.

Resistance Standards. Standards of resistance are made in various forms according to the standard which they

represent and the uses to which they are to be put.   The
standards of reference, such as the standard ohm, are made
on special bobbins, with elaborate arrangements for maintaining
them at a constant temperature.   Great care is also taken
with the means of making connexions.   The resistance itself
is usually made of an alloy like platinoid or manganin or
platinum-silver and is wound double—back on itself—after
the manner of the non-inductive coil shewn by Fig. 90.
Moreover it usually consists of *two* coils in parallel, one of
much finer wire than the other.   The fine wire coil is the
one which is altered in making the final adjustments of the
standard.

In electrical engineering practice it is common to have
low resistance standards (e.g. 0·1 and 0·01 ohm) for use in

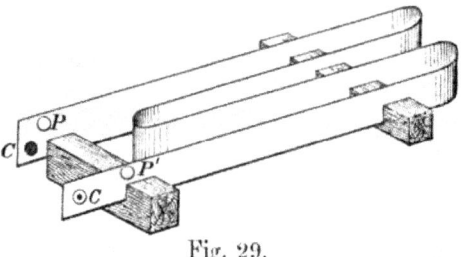

Fig. 29.

current measurement, such as that described at the end of
this chapter in the paragraph on the potentiometer method.
These standards must be made so that a large current may be
passed through them without appreciable heating.   A 0·1 ohm
standard must carry 10 amperes without heating ; a 0·01 ohm
must carry 100 amperes, and so on, for the particular uses
of the standards, the object being as a rule to measure the
current in the resistance by determining the drop in volts
through it.   This is usually done with the potentiometer.

These *Current Resistances* as they are sometimes called
are usually made in the form of thick manganin wire, or, in
the case of the lower resistances, of manganin strip with a
large cooling surface.

They have two pairs of terminals, one pair called the *current terminals* and the other pair the *potential terminals*. These are shewn by CC and PP' respectively in Fig. 29, which is an illustration of a 0·01 ohm standard resistance to carry 100 amperes. The resistance of the standard is that between the *potential terminals PP'*, therefore it is very necessary that the connexions are made properly.

MEASUREMENT OF RESISTANCE.

All methods in common practice for the measurement of resistance consist in comparing the unknown resistance with a known standard. Resistance can always be determined by measuring the difference of potential across the ends of the unknown when a known current is passing through it. Such methods are crude however, inasmuch as there are two possible sources of error, one in measurement of the current and the other in the measurement of the difference of potential. The method involves the use of *two* measuring instruments—and few ammeters or voltmeters would be sufficiently accurate, for anything but approximation.

Resistance measurement methods are divided into three classes; methods for high resistance, for low resistance, and for ordinary resistance. Resistances between 1 and 10,000 ohms are termed ordinary resistances; above 10,000 ohms they are called *High*; below 1 ohm they are called *Low*, and the methods commonly adopted for the measurement of one class are not suitable for measurement of the others.

High Resistance: method of substitution. High resistances are generally measured by the method known as the method of substitution. This consists in substituting a known standard in place of the unknown, in a circuit consisting essentially of a delicate galvanometer and a battery, and determining the deflexions produced in each case. Fig. 30 is a diagram of the connexions.

The unknown resistance R is firstly put in the circuit and the deflexion of the spot of light on the scale is noted—right and left of zero. The mean deflexion can be represented by D_R. It should be pointed out that the current in the circuit will be necessarily small since R is a high resistance. Therefore it is necessary to have a sensitive galvanometer—and therefore one of many turns of wire having in itself a high

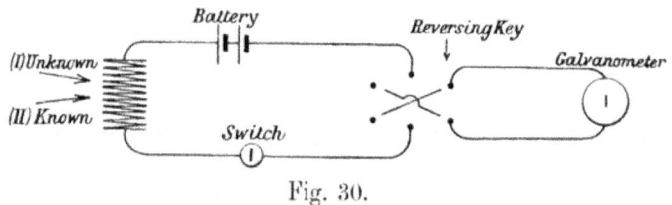

Fig. 30.

resistance. It can also be noted that a mirror galvanometer must be used since sensitiveness is required, and that the deflexions of the spot of light are proportional to the currents producing them. Let G be the resistance of the galvanometer. This is always supplied with the galvanometer. If E denotes the E.M.F. of the battery, then the current in the circuit will be found by

$$C_R = \frac{E}{R + G + b},$$

where b is the resistance of the battery. Now R is high, therefore b may be safely neglected, and the current C_R in this circuit will be given by

$$C_R = \frac{E}{R + G}.$$

Now the known resistance K is substituted for R, and deflexions are taken as before. Let the mean deflexion be D_K. The current in the circuit in this case is

$$C_K = \frac{E}{K + G}.$$

Now the deflexions are proportional to the currents,

$$\therefore \ \frac{D_R}{D_K} = \frac{C_R}{C_K} = \frac{\dfrac{E}{R+G}}{\dfrac{E}{K+G}} \ .$$

But E is the same in both cases,

$$\therefore \ \frac{D_R}{D_K} = \frac{K+G}{R+G},$$

whence R (the unknown)

$$= \frac{D_K(K+G)}{D_R} - G,$$

and D_K, D_R, K, and G are all determined. Thus R is calculated.

Now this method as described above is quite excellent provided that R and K are of the same order of resistance. But if they are very different—if R is 100 times K for example—then the deflexions resulting will be so different that it will be frequently impossible to get them both on the scale. Or even if one could, one of them would be very very small, and the other very great; and it is not advisable to work as a rule with either the one or the other. An error of reading in the case of a small deflexion produces a big error in the result; and the large deflexions are not strictly proportional to the deflecting currents. In practice the deflexions are generally adjusted as far as possible so that the spot does not extend beyond 5 inches on each side of the zero at a scale distance of 40 inches from the galvanometer.

Again, in the measurement of high resistance it is not usual, nor economical, to have a large number of standards. Consequently some method must be employed so that a very high resistance can be compared against the comparatively low resistance of the standard. The standard will be in all probability 1,000,000 ohms—called a megohm—or it might be one-tenth of this.

Principle and use of the Shunt.   The method employed is to connect up to the galvanometer a resistance coil whose resistance is a definite fraction of that of the galvanometer, say $\frac{1}{9}$th.   This resistance is merely connected to the terminals of the galvanometer, quite independently

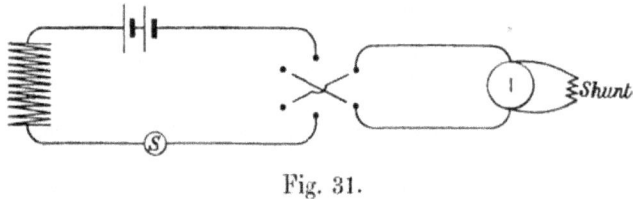

Fig. 31.

of the remainder of the circuit which remains as described above.   The galvanometer is now said to be *shunted*, and this resistance is called the *shunt*.

Fig. 31 illustrates the connexions for the method when a shunt is used.

The combined resistance of the galvanometer and the shunt, the former denoted by G, and the latter by S, will be

$$\frac{1}{\frac{1}{G}+\frac{1}{S}} = \frac{G \times S}{G + S},$$

and this will always be less than G or than S.

Now the current of the main circuit will divide through G and S.   It will split into $(G + S)$ parts and S of these will go through the galvanometer and the remainder through the shunt.

Thus if S be made small only a very small fraction of the total current will pass through the galvanometer.

Again the fraction of the current in the galvanometer must be $\frac{S}{G + S}$ of the total current in the circuit whilst $\frac{G}{G + S}$ of the current passes through the shunt.

As a result of this the deflexion which the same total

current would have produced if it passed through the galvanometer would be $\dfrac{S+G}{S}$ times the deflexion produced when the shunt S was used.

A definite case may be taken, when G is 9000 and S is 1000 ohms. The current of the main circuit will split into 10 parts, 9 of which will go through the shunt and 1 through the galvanometer.

That is to say, the fraction of that current which passes through the galvanometer

$$= \frac{S}{G+S} = \frac{1000}{10000} = \frac{1}{10}.$$

Therefore it follows that had the total current passed through the galvanometer the deflexion produced would have been 10 times *greater*.

Thus it may be stated as a general rule that if a shunt S be used, then the deflexion obtained will be $\dfrac{S}{G+S}$ of the deflexion which would have been obtained without the shunt.

Now in practice, when the comparisons are being made readings are taken with shunts which bring the deflexion to a reasonable order. These deflexions are then expressed in terms of what they would have been without the shunt. This is simply done.

Equivalent deflexion without shunt

$$= \text{deflexion with shunt} \times \frac{G+S}{S}.$$

The expression $\dfrac{G+S}{S}$ is called the *multiplying power* of the shunt.

Another point which will at once occur to the reader is that the combined resistance of G and S being less than G the total current in the circuit will be altered by the introduction of the shunt. As a matter of fact the combined resistance is usually so small compared with the known or unknown high resistances that the error produced by neglecting it entirely will not be greater than 1 part in 1000. And thus it is usual to work always with a shunt and to neglect the resistance of galvanometer and shunt. Thus the expression for the unknown resistance R becomes

$$R = \frac{D_K \times K}{D_R},$$

where D_K and D_R are *the equivalent deflexions without shunts* when K and R respectively are the resistances in the circuit.

The *common* form of shunt consists of a box with three coils, being of resistance $\frac{1}{9} G$, $\frac{1}{99} G$, and $\frac{1}{999} G$ respectively. Each galvanometer must have its own shunt box.

With these shunts the currents in the galvanometer would be $\frac{1}{10}$, $\frac{1}{100}$ and $\frac{1}{1000}$ of the total current in the circuit respectively.

The Universal Shunt. A form of shunt box known as a *universal shunt* has been devised, and this can be used for any galvanometer of any resistance.

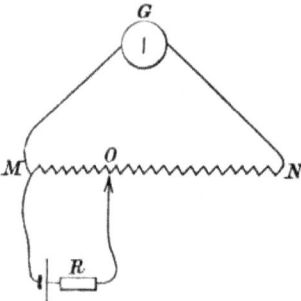

Fig. 32.

The principle of the universal shunt is illustrated by Fig. 32. A high resistance MN is connected to the galvanometer terminals, and the "line" connexions are made from M and some other point O on MN. The point O may be made at a series of positions between M and N, the resistance MO being a definite fraction of the resistance of MN.

The current strength in a shunted galvanometer,

$$c = C\,\frac{S}{S+G},$$

where C is the total current and S the shunt resistance. Therefore in this case

$$c = C\,\frac{MO}{MO + \{(MN - MO) + G\}}$$

$$= C\,\frac{MO}{MN + G},$$

where MO and MN are the resistances between these respective points.

It will be seen therefore that the fraction of the total current in the galvanometer will depend upon the ratio of $\dfrac{MO}{MN}$.

For example, in one case, let $MO = \frac{1}{3}\,MN$,

$$\therefore c_1 = C_1\,\frac{\frac{1}{3}\,MN}{MN + G}.$$

In a second case, let $MO = \frac{1}{10}\,MN$.

$$\therefore c_2 = C_2\,\frac{\frac{1}{10}\,MN}{MN + G}.$$

$$\therefore \frac{c_1}{c_2} = \frac{C_1\,\dfrac{\frac{1}{3}\,MN}{MN + G}}{C_2\,\dfrac{\frac{1}{10}\,MN}{MN + G}}.$$

$$\therefore \frac{c_1}{c_2} = \frac{C_1\,\frac{1}{3}}{C_2\,\frac{1}{10}}.$$

It is thus seen that the fraction of the total current in the galvanometer is *independent of the galvanometer resistance*, and therefore the universal shunt can be used with any galvanometer.

The universal shunt is generally made on the "dial" pattern with a contact arm which can be turned about A (Fig. 33) to make contact at O_1, O_2, &c. The method of connecting is shewn. The resistances of the several portions are as follows :

$$MO_1 = \tfrac{1}{3} \ MN,$$
$$MO_2 = \tfrac{1}{10} \ MN,$$
$$MO_3 = \tfrac{1}{30} \ MN,$$
$$MO_4 = \tfrac{1}{100} \ MN,$$
$$MO_5 = \tfrac{1}{300} \ MN,$$
$$MO_6 = \tfrac{1}{1000} \ MN.$$

The galvanometer is never used absolutely *without shunt*, for when the arm is connecting AN the whole shunt MN is in

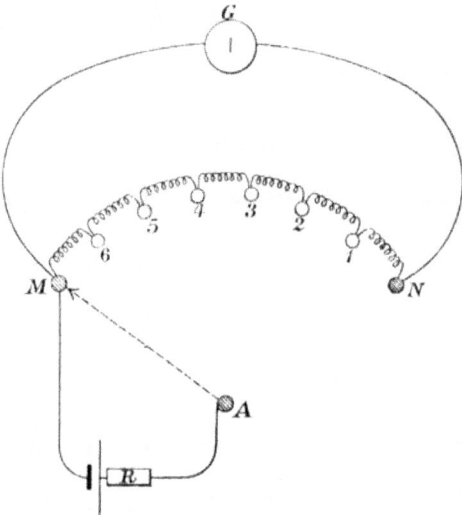

Fig. 33.

with the galvanometer. Then it is taken that the whole current is passing. When the arm is across AO_1, then $\tfrac{1}{3}$ of this is passing, and so on.

The deflexion obtained multiplied by 3, or 10, or 30, &c., may be taken as the *equivalent deflexion without shunt*—although it will really be the equivalent deflexion with the whole shunt MN in. The fractions marked on the box represent the fraction of the whole current going through the galvanometer and not the fraction of the resistance of MN. Thus O_1 will be marked $\frac{1}{3}$; O_2 will be marked $\frac{1}{10}$; O_3 $\frac{1}{30}$; O_4 $\frac{1}{100}$; O_5 $\frac{1}{300}$; O_6 $\frac{1}{1000}$; and N, 1.

It should be pointed out that the total current C will not remain absolutely the same as the position of the contact arm is changed, since the combined resistance of galvanometer and shunt will not be constant. However, since this is chiefly used for the measurement of *high* resistances the error will only be very slight, and will rarely exceed 1 part in 1000.

This method of substitution is the method employed for the measurement of insulation resistance. The insulation resistance of a coil of cable, for example, is determined by immersing the coil in a water vat—both ends of the cable being out of the water. The resistance between one end of the copper wires of the cable and a plate of copper immersed in the middle of the coil is then measured. This is usually done at a high E.M.F.—at least twice the E.M.F. for which the cable is to be used at any time. The result is expressed in megohms per mile, the insulation resistance being inversely proportional to the length of the cable. Hence it is necessary to know the length of the cable in the vat. This measurement is described on page 180.

Low Resistance Measurements. Low resistances are generally measured by a method known as the fall of potential method, which consists in comparing the E.M.F.s across a known and the unknown low resistance when equal currents are passing through them.

Fig. 34 is a diagrammatic illustration of the connexions of the circuits required. A cell E is connected, through a reversing key, to a variable resistance VR (for adjustment of the current in the circuit), the known resistance K, and an unknown resistance R. This forms a simple series circuit and the current strength will be equal therefore in all parts of it. If the strength of this current be denoted by C, it follows that the difference of potential across K will $= C \times K$; and the difference of potential across R will be $C \times R$.

A high resistance galvanometer G is arranged so that it can be connected across the terminals of K or of R at will by means of a change key CK.   This galvanometer will act like

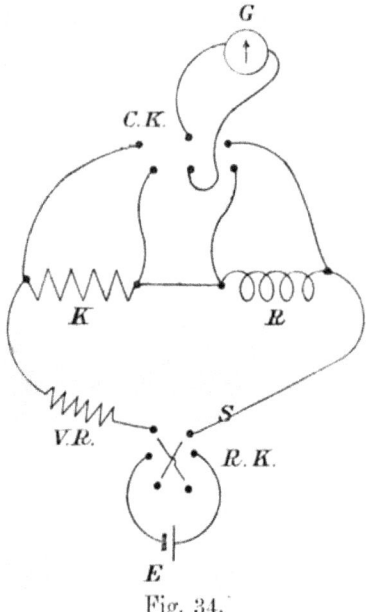

Fig. 34.

a voltmeter and therefore the deflexions produced will be proportional to the differences of potential across its terminals.

The deflexion when it is connected across K may be denoted by D_K; and when connected across R by D_R; deflexion being understood to be *mean* deflexion in all cases.

Now :

$$\frac{\text{Difference of potential across } K}{\text{Difference of potential across } R} = \frac{C \times K}{C \times R} = \frac{K}{R},$$

C being equal in both.

And, since $\dfrac{\text{difference of potential across } K}{\text{difference of potential across } R} = \dfrac{D_K}{D_R}$,

$$\therefore \frac{D_K}{D_R} = \frac{K}{R}.$$

$$\therefore R = \frac{D_R \times K}{D_K}.$$

This method admits of considerable accuracy, but the galvanometer resistance must be very high and great care must be taken in making the connexions across K and R. If the adjacent ends of K and R be separated by a piece of wire then separate wires must be run from each end to the galvanometer terminal, for it must be remembered that K and R are low resistances and probably less than the resistance of the connecting wires.

A number of values of R should be determined by varying the variable resistance and so varying the current in the main circuit.

Comparison of Methods. The method adopted for high resistance measurement could not be used for low resistances, since the substitution of one low resistance for another in a circuit which would be necessarily of high resistance would not produce any appreciable alteration in the strength of the current in the circuit, and consequently in the deflexions produced.

Conversely, the fall of potential method does not lend itself to high resistance measurement, since the resistance of the galvanometer in use as a voltmeter would be less than the resistance across which it was connected. Thus the total resistance would be considerably reduced and no simple expression could be given for the relationship of the deflexions and the resistances.

Ordinary Resistance Method : Wheatstone Bridge. Ordinary resistances, that is to say between 1 ohm and

10,000 ohms, are measured usually by a method known as the Wheatstone Bridge method, the principle of which may be described simply as follows.

  ABC, and *ADC*, in Fig. 35, are two resistances connected up in parallel with each other and to a cell. A current passes which divides at *A*, part passing along *ABC* and part along *ADC*, and these unite again at *C*. The strengths of the currents in the two parts will be in inverse ratio to the

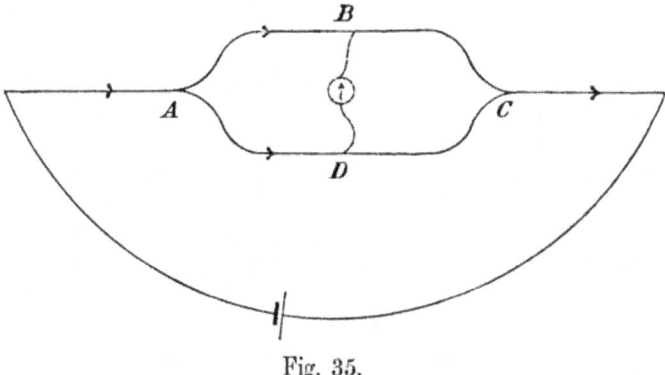

Fig. 35.

resistances. Now the difference of potential across *ABC* must be equal to that across *ADC*. Along *ABC* there will be points at every potential between that of *A* and that of *C*; and there will be similar points along *ADC*. Therefore every point along *ABC* will have a corresponding point along *ADC* at which the potential will be the same as its own. If, therefore, a galvanometer be connected to these corresponding points no current can pass through it.

  Now the potential of any point on *ABC* will depend upon the resistance between that point and *A* ; and consequently on the resistance between that point and *C*. If the point *B* be considered, the difference of potential between *AB* will be to that between *AC* as their resistances.

$$\therefore \quad \frac{\text{Difference of potential across } AB}{\text{Difference of potential across } AC} = \frac{\text{Resistance } AB}{\text{Resistance } ABC}.$$

$$\therefore \quad \text{Difference of potential across } AB = \frac{\text{Resistance } AB}{\text{Resistance } ABC}$$

$$\times \text{ difference of potential across } ABC.$$

Clearly a point D on the resistance ADC will have the same potential as the point B when

$$\frac{\text{Resistance } AD}{\text{Resistance } ADC} = \frac{\text{Resistance } AB}{\text{Resistance } ABC},$$

since the difference of potential across $AD = \dfrac{\text{Resistance } AD}{\text{Resistance } ADC}$

$$\times \text{ difference of potential across } ADC,$$

and the difference of potential across ADC is equal to that across ABC.

Similarly it follows that

$$\frac{\text{Resistance } BC}{\text{Resistance } ABC} = \frac{\text{Resistance } DC}{\text{Resistance } ADC}$$

when there is no difference of potential between B and D.

Therefore it follows that

$$\frac{\dfrac{AB}{ABC}}{\dfrac{BC}{ABC}} = \frac{\dfrac{AD}{ADC}}{\dfrac{DC}{ADC}}.$$

$$\therefore \quad \frac{AB}{BC} = \frac{AD}{DC} \text{ when } B \text{ and } D \text{ are corresponding points.}$$

This may be considered in other ways, but the conclusion arrived at will always be the same. This conclusion is applied to the measurement of ordinary resistances. The following methods differ only in practical form.

Metre Wire Bridge. The metre wire bridge consists essentially of a length of platinoid wire, 1 metre long, of uniform area of cross section, stretched along a base board.

A metre scale, divided from each end, is fixed adjacent to
this wire, and some form of contact maker (known as a
jockey) is arranged so that it can be moved along to any
part of the wire.   This wire is connected to terminals by
heavy bars of copper of negligible resistance, and additional
terminals are provided after the manner illustrated in
Fig. 36.

This is used to compare the resistances, and is connected
up as shewn by the figure.   The lettering used corresponds
to that in Fig. 35.   The resistances to be compared, r_1 and r_2,

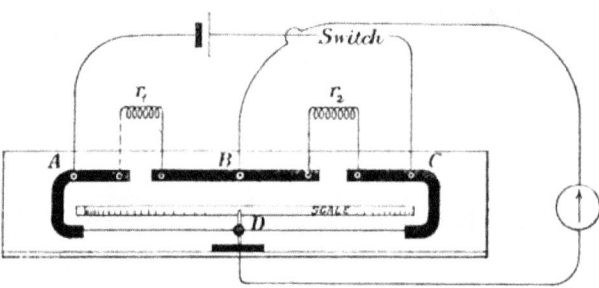

Fig. 36.

form the arms AB and BC of the "Bridge."   The cell is
connected to A and C.   The galvanometer is connected to
B and to the *jockey*.   The jockey is moved along the wire
until a point is found where no deflexion of the galvanometer
needle is produced.   This point is D, and AD and DC will
give the remaining arms of the Bridge.   The point D is easily
determined.   When the jockey is one side the galvanometer
needle will be deflected to the right : when on the other side
the deflexion will be to the left.

Now when a balance is obtained there will be no difference
of potential between B and D.

$$\therefore \frac{AB}{BC} = \frac{AD}{DC}.$$

But the lengths AD, and DC, of the wire will be proportional to their resistances.

$$\therefore \frac{\text{Length } AD}{\text{Length } DC} = \frac{AB}{BC} = \frac{r_1}{r_2}.$$

And if either r_1 or r_2 be known the other can be calculated.

The point D at which the balance is obtained will *not* vary with the strength of the current of the cell, since all variations of potential due to varying currents will be proportional.

The method admits of some errors. Firstly the wire may not be uniform. To correct for this the positions of r_1 and r_2 should be interchanged and a new position found for D. *Means* should be taken. Secondly, unless r_1 and r_2 are approximately of the same order D will be near one end or other of the wire. Thus errors of reading the position of D on the scale would become greater in proportion. Hence the method is only recommended for comparison of nearly equal resistances.

Post Office form of Wheatstone Bridge. The method adopted by the Post Office for the measurement of ordinary

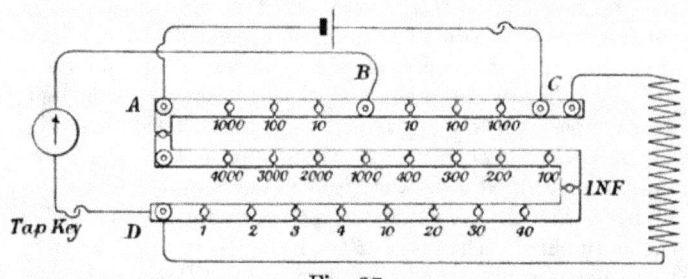

Fig. 37.

resistances is merely a modification of the metre wire bridge. Fig. 37 illustrates the principle of the Post Office method. This is an ordinary series box but with an extra set of coils known as the *ratio arms*, and terminals as illustrated. The lettering on the figure corresponds to that used in the two preceding diagrams. In use, the cell is connected to A and C through a tapping contact key; the unknown resistance is connected to C and D. Thus there are the two paths for the current—one through ABC and the other through ADC. The galvanometer is connected to B and D, through a tapping key.

Now the ratio arms can have six simple ratios varying from 100 : 1 to 1 : 100, since there are three coils on each arm of 1000, 100, and 10 ohms respectively.

Firstly the resistances of AB and BC are made 10 ohms each—that is a ratio of 1. AD is made "infinity" by taking out the "infinity" plug. The battery key is put down and the galvanometer key is tapped smartly. If there is no deflexion then CD must be infinity, since $AB:BC::AD:DC$. If there is a deflexion then its direction is noted, and the experimenter knows that whenever the deflexion produced is in that direction the resistance in AD is too great.

The infinity plug is put back again and 1 ohm is taken out. If the deflexion is in the opposite direction, then 1 ohm is too little. Now a balance will be obtained when AD is equal to DC, since AB has been made equal to BC. Thus the resistance of AD is varied until a correct balance is obtained or until two resistances differing only by one ohm are found, one of which causes a deflexion in one direction and the other in the other direction. It may be supposed, for example, that when AD is 8 ohms the deflexion is to the left and when AD is 9 ohms the deflexion is to the right. Then the required resistance is between 8 and 9 ohms.

Then the ratio is altered. AB is made 100 and BC is left at 10 ohms. Thus $AB:BC=100:10=10:1$. Therefore if a balance is obtained by adjustment of AD, the resistance of AD will be 10 times that of DC, the unknown. It may be supposed that with AD of 86 ohms the deflexion is left; of 87 ohms the deflexion is right. Then the required resistance of DC is between 8·6 and 8·7 ohms.

Then the ratio is further altered. AB is made 1000, BC being left at 10 as before. Now $AB:BC=100:1$,

AD is again adjusted and it can be supposed that a balance is obtained—that is, no galvanometer deflexion is produced on putting down keys—when AD is 863 ohms.

Now $\qquad AB : BC :: 100 : 1,$

$\therefore \quad AD : DC :: 100 : 1,$

$\therefore \quad 863 : DC :: 100 : 1,$

$\therefore \quad DC = \frac{863}{100} = 8\cdot63$ ohms.

This method is very simple and a little practice will enable any experimentalist to find a balance very quickly. It is seen that the method differs from the metre wire bridge only

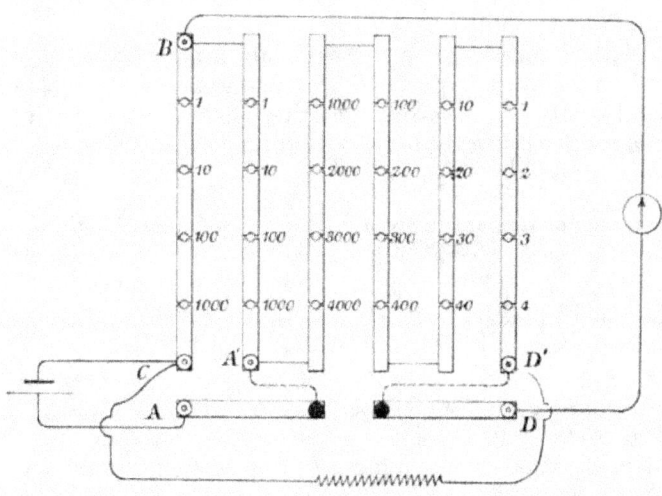

Fig. 38.

in the manner of adjustment. It must be pointed out that an exact balance is not always to be obtained—but the resistance required can always be calculated by reading the right and left deflexions produced when AD is adjusted to the two nearest resistances. It will be seen by the resistances of the ratio arms any resistance between 1 and 1000 ohms can be determined to within 0·1 %.

A recent form of Post Office Box is illustrated by Fig. 38

which represents the top of the box. The lettering corresponds to the previous figures, and the tapping keys are already connected up in the box—the connexions being shewn by dotted lines.

The battery key should always be put down before the galvanometer key. This is done to get rid of self-induction effects in the unknown resistance. If DC be an inductive coil, and the bridge be balanced as described above; then if the galvanometer key be put down first and the battery key second it will be found that a deflexion is produced. This is due to induction. The other arms of the bridge are non-inductive coils—wound back on themselves as described previously.

Methods in which measurements are made by adjusting until no deflexion is produced are known as *Null* methods. They have the advantage over other methods that the accuracy or inaccuracy of a galvanometer reading does not enter into the result.

Internal Resistance of a Battery: Voltmeter Method.

There are many methods by means of which the internal resistance of a cell or a battery of cells may be determined. The following method will appeal to the electrical engineering student, although it admits of errors.

The cell or battery, as the case may be, is connected up to a voltmeter as shewn in Fig. 39a. The reading gives the E.M.F. of the cell, it being understood that the E.M.F. of a cell is measured by the difference of potential between its terminals when it is on "open circuit," that is to say doing no work. This E.M.F. may be denoted by E.

A known resistance r is connected to the cell terminals and a current will pass round the circuit. The current strength will depend upon the E.M.F. acting and upon the *total* resistance. If it be denoted by C, then

$$C = \frac{E}{r+b},$$

where b is the internal resistance of the cell or battery. Now in this expression there are two unknowns—viz. C and b; consequently neither can be calculated with the data already found.

However C may be found by connecting the voltmeter to the ends of the resistance r as illustrated by Fig. 39c. The current in this resistance will be equal to the difference

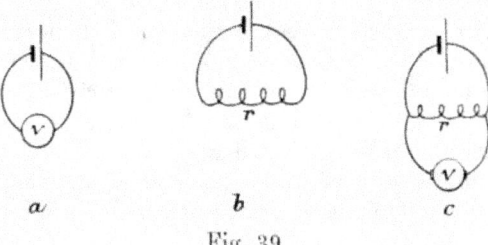

a b c

Fig. 39.

of potential across its ends divided by r. Thus if e is the difference of potential across r, then

$$\text{the } current \text{ in the resistance } r = \frac{e}{r}.$$

But the circuit shewn in Fig. 39b is a simple circuit and the current strength is the same at all parts of it. Hence it follows that

$$\text{the current strength in the resistance } r = C,$$

$$\therefore \; C = \frac{e}{r} = \frac{E}{r+b}.$$

Hence $$(b+r)e = E \times r,$$

$$\therefore \; eb + er = Er,$$

$$\therefore \; b = \frac{Er - er}{e} = \frac{(E-e)r}{e}.$$

It may be looked at in another way, for $(E-e)$ represents the E.M.F. necessary to urge the current C through the resistance b of the cell.

$$\therefore \ C = \frac{E - e}{b},$$

$$\therefore \ \frac{e}{r} = \frac{E - e}{b},$$

$$\therefore \ b = \frac{(E - e)\,r}{e},$$

which is identical with the expression above.

In making a measurement in this way it is necessary to work quickly. The resistance r should not be connected to the cell for any time longer than is absolutely necessary to take the reading of the voltmeter. The object of this is to prevent polarisation, which would alter the results completely. Polarisation not only increases the internal resistance of a cell but also sets up a back E.M.F. which reduces the working E.M.F.

If, for example, the back E.M.F. be denoted by E'', then the current in the circuit would be

$$C = \frac{E - E'}{r + b + b'},$$

where b' denotes the additional internal resistance caused by the polarisation of the cell. Hence the necessity of keeping down polarisation. To this end the resistance r should be chosen as high as practicable, that is to say it should be high but not so high that the resistance of the cell becomes incomparable with it.

Now the lower the value of r the lower is the value of e and *vice versa*. Hence if r is practically zero it follows that all the work will be done inside the cell, and that e will be zero.

Again, if r be high then e will be high, until finally $e = E$. Thus it is that in performing the experiment r should be adjusted so that it is as high as possible, the adjustment being made so that there is a measurable difference between e and E, and no more.

Arrangement of Cells in Series and Parallel. In determining the internal resistance of a number of cells, that

is to say of a battery, the student will find that it is dependent upon the arrangement of the cells.   Firstly it is general
to have batteries composed of exactly similar cells, and consequently with approximately equal individual resistances and
E.M.F.S.

The total resistance of a number of cells follows the rules
governing the resistance of a number of conductors.   In
series, the total resistance is the sum of the individual resistances.   Thus if there are n cells of equal resistance b, in
series the combined resistance B will $= n \times b$.

The same cells in parallel would have a resistance

$$B = \cfrac{1}{\cfrac{1}{b_1} + \cfrac{1}{b_2} + \dots + \cfrac{1}{b_n}},$$

but as the cells have equal resistances,

$$\therefore\ B = \frac{b}{n},$$

where b is the resistance of any one of them.   Hence parallel-
ing cells reduces the total internal resistance of the battery.

The student will know both by experience and by his
elementary reading of the subject that the E.M.F. of a number
of cells in series is the sum of the individual E.M.F.s ; whilst
the total E.M.F. of a number of equal cells in parallel is the
same as the E.M.F. of any one.

Hence with a battery of n equal cells, each having an E.M.F.
E and an internal resistance b,

Connexion	Total E.M.F.	Total Internal Resistance
Series	$n \times E$	...... $n \times b$
Parallel	E	...... $b \div n$

Compounds of series and parallel may be made.   A number
of cells may be divided into two equal sets ; each set is connected up in series and the two sets paralleled.   This arrangement will give a total E.M.F. of $\frac{n}{2} \times E$ and a total internal
resistance of $\frac{n}{4} \times b$.   The reader can easily verify this.

The idea of arranging a given number of cells in different ways is to get as great a current as possible—or necessary—through a given external circuit. It is shewn that the greatest current is produced in a given circuit when the cells are so arranged that the total internal resistance of the battery is as nearly equal as possible to the external resistance.

It may be supposed that there are n cells in the battery, the E.M.F. of each being e and the internal resistance of each b. The external resistance may be denoted by r.

Let there be L sets of K cells in series. That is to say, there are L groups of cells coupled in parallel, each group consisting of K cells in series.

Then $$K \times L = n.$$

The resistance of K cells in series $= K \times b$.

The resistance of L groups in parallel $= \dfrac{Kb}{L} =$ the total internal resistance.

The E.M.F. of K cells is series $= K \times e =$ the total E.M.F. of the battery.

Now   current $C = \dfrac{\text{Total E.M.F}}{\text{Total resistance}}$

$$= \frac{Ke}{\dfrac{Kb}{L} + r} = \frac{Ke}{\dfrac{Kb}{L} + \dfrac{Lr}{L}}$$

$$= \frac{KLe}{Kb + Lr}.$$

But $KL = n, \; \therefore \; C = \dfrac{ne}{Kb + Lr}.$

Now $n \times e$ must be constant for n cells of E.M.F. e, therefore for a maximum current $Kb + Lr$ must be as small as possible.

Now the product of Kb and $Lr = KLbr = nbr$, and nbr must also be a constant.

When the product of two numbers is a constant the sum of those numbers is least when they are equal.

For example, $5 \times 20 = 100$; $10 \times 10 = 100$; $25 \times 4 = 100$; $40 \times 2 \cdot 5 = 100$; $50 \times 2 = 100$; $100 \times 1 = 100$. Therefore the least value for $Kb + Lr$ will be given when $Kb = Lr$.

That is, when $$\frac{Kb}{L} = r,$$

but $\frac{Kb}{L}$ is the total internal resistance, and r is the external resistance.

Hence it is proved that a maximum current is obtained through a given resistance with a given number of cells when they are so arranged that the internal resistance is as nearly equal as possible to the external resistance.

Efficiency of a Battery. Now although this arrangement produces the maximum current it does not give a maximum of efficiency. The efficiency of a battery under working conditions is the ratio of the useful work to the total work ; or of the work done in the outside circuit to the whole work done in the system. For a given time therefore the total work done is

$$E \times C \times t,$$

where E is the E.M.F. of the battery, C the total current strength, and t the time.

In the outside circuit this will be $e \times C \times t$, where e is the difference of potential across the outside resistance when the current C is flowing.

Thus the efficiency of the system will be

$$\frac{e \times C \times t}{E \times C \times t} = \frac{e}{E},$$

since the time will be the same inside and outside, and the total current C will also be the same.

Thus as e is smaller the efficiency will be less, and when $e = E$ the efficiency will be a maximum, viz. $100°/_{\circ}$. If the battery resistance be equal to resistance of the outside circuit, then e will be half of E. Therefore the efficiency will be only $50°/_{\circ}$. If the battery resistance be greater than the outside resistance then e will be less again. In short the ratio of e to E will be the same as the ratio of the outside resistance to the total resistance. Hence the efficiency of the system is

$$\frac{e}{E}, \text{ which is the same ratio as } \frac{r}{R},$$

where r is the outside and R the total resistance. Thus for a *maximum efficiency* r must be as nearly equal to R as is possible.

This applies equally to the case of a dynamo and its outside circuit. The internal resistance must be kept down so that a minimum of work is done in the inside. And this is the point to be considered at all times for economic reasons.

Measurements with the Potentiometer. Resistances, currents, and E.M.F.s may be compared and measured by means of the *potentiometer*. As its name implies, the apparatus primarily measures E.M.F.'s ; but it may be used for

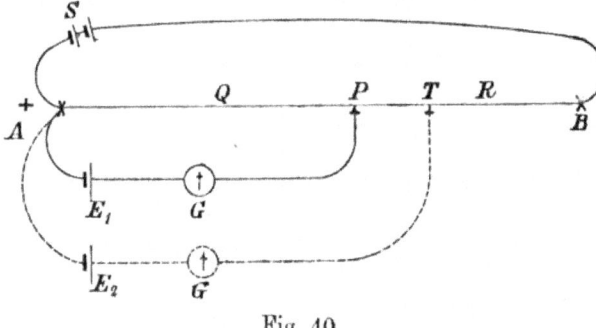

Fig. 40.

resistance and current measurements, and no testing laboratory can be considered complete if it has not some form of potentiometer in its equipment.

The principle may be explained by a simple experiment for the comparison of the E.M.F.s of two cells.

AB, in Fig. 40, is a long *uniform* thin wire, the resistance of which is strictly proportional to its length. A pair, say, of secondary cells are connected to AB. Thus a current will be maintained along the wire and there will be a difference of potential between A and B. It is essential that the strength of this current should remain constant so that the difference of potential between A and B is constant and that is why secondary cells are used to maintain it.

The cells whose E.M.F.s are to be compared are E_1 and E_2. E_1 is connected up as shewn, its positive terminal being connected to A, which is also connected to the positive of the battery S. The negative is connected to a galvanometer, and the other terminal of the galvanometer is connected to some form of movable contact and in this manner with the wire AB.

Now if the difference of potential between A and B be greater than the E.M.F. of E_1, then since the point A must be at the same potential as the positive terminal of E_1 it must follow that some point along the wire between A and B must be at the same potential as the negative terminal. If P be that point, then when contact is made with it there will be no flow of current through the galvanometer since both terminals are at the same potential. Under such conditions if contact be made at Q, then since Q will be at a higher potential than the negative terminal of E_1 a current will flow from Q through the galvanometer. If contact be made at R, since R will be at a lower potential than the negative terminal of E_1, a current will flow from the terminal through the galvanometer to R.

Hence right and left deflexions are obtained and the point of balance P may easily be found.

Then E_2 is joined up similarly—shewn by dotted lines— and a "balance" is obtained at the point T.

Now it will be seen that

$$\frac{\text{E.M.F. of } E_1}{\text{Difference of Potential between } AB} = \frac{\text{Current strength in wire} \times \text{Resis. of } AP}{\text{Current strength in wire} \times \text{Resis. of } AB}.$$

$$\therefore \frac{E_1}{\text{Difference of Potential between } AB} = \frac{\text{Resis. } AP}{\text{Resis. } AB} = \frac{\text{Length } AP}{\text{Length } AB}.$$

But this would merely compare E_1 with the unknown potential difference across AB.

Similarly with E_2,

$$\frac{E_2}{\text{Difference of potential between } AB} = \frac{\text{Length } AT}{\text{Length } AB}.$$

Now since E_1 and E_2 have been compared with the same

E.M.F.—viz. the difference of potential between A and B, they may be compared with one another.

$$\therefore \frac{E_1}{E_2} = \frac{\text{Length } AP}{\text{Length } AT}.$$

If either E_1 or E_2 be known the other can be determined. One of these can be a Clark standard cell. A resistance should be inserted between one terminal of the cell under test and the galvanometer. This should be about 10,000 ohms and it serves to prevent the galvanometer needle from being too violently deflected, and what is more important from the test point of view prevents any appreciable current passing through the cell, for polarisation in either direction must be prevented ; and furthermore the terminal potential difference of the cell would be altered by a current passing.

For the above comparison a simple potentiometer consists of some six metres of fine platinoid wire stretched in six lengths on a board. A double reading scale and a movable contact maker or "jockey" complete the apparatus.

Principle of Central Station Potentiometer. In testing work a different method is adopted—although the

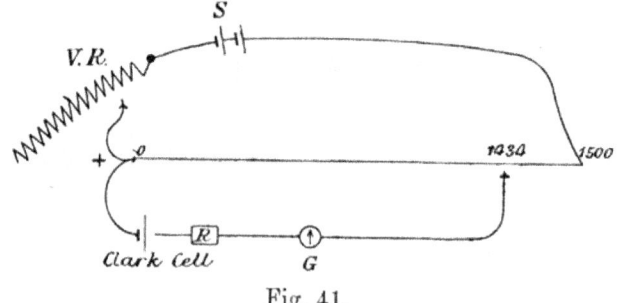

Fig. 41.

underlying principle is just the same. The essential difference lies in "standardising the wire" at the outset by means of a Clark cell.

The wire is made 1500 units long (though one can always read to a tenth of a unit) and is connected up as shewn in Fig. 41. A variable resistance, v.r., is included in the secondary cell circuit. A Clark cell is connected with the positive terminal of the wire and through a resistance and the galvanometer to the wire. The contact is made at 1434 units (for the e.m.f. of the Clark cell is 1·434 volts at 15° C.) if the temperature be 15° C., or at a corresponding point if the temperature be different. Then the variable resistance v.r. is adjusted *until the balance is obtained when the jockey is at* 1434.

The wire is then standardised. 1000 units of wire under these conditions represents a potential difference of 1 volt. Now if any cell is to be tested it is connected up as the Clark cell is shewn to be and the *jockey* is now moved until a balance is found. The variable resistance *must not be altered* once the wire has been standardised with the Clark cell. All other adjustments are made with the jockey. When a balance has been found the position of the jockey indicates volts × 1000. Hence the jockey reading ÷ 1000 gives the e.m.f. of the cell. The potentiometer scale thus becomes " direct reading."

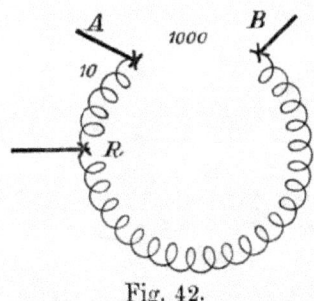

Fig. 42.

It will be noted that there is the apparent limit of 1·5 volts, but if it is wished to read between 1·5 and 3·0 volts for example, the wire is standardised with the Clark cell when the jockey is at $\frac{1434}{2}$, viz. 717 units. Thus the reading of the jockey must be doubled, then ÷ 1000 and the result is given in volts.

For higher E.M.F.s it is usual to have *ratio resistances*. Thus if it was wished to measure an E.M.F. of about 100 volts a non-inductive resistance of say 1000 ohms would be joined up to the 100 volt terminals. If two points on this resistance be chosen, between which there is a resistance of 10 ohms, then the difference of potential between these will be $\frac{1}{100}$ that between the main terminals. This is illustrated by Fig. 42. AB is say 1000 ohms. AR is $\frac{1}{100}$ of this. Thus if 100 volts be applied to AB the potential difference between AR will be 1 volt and *this can be measured on the potentiometer*. The result is multiplied by the ratio of $\dfrac{AB}{AR}$ and the E.M.F. applied to AB is determined.

Different ratio terminals are given on the special ratio resistances made for the potentiometer.

Measurement of Current by Potentiometer. This consists merely in measuring the difference of potential

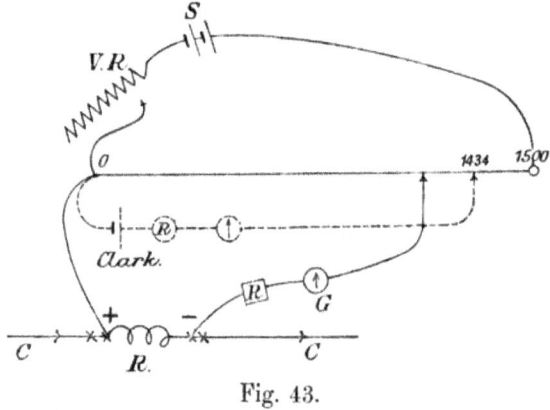

Fig. 43.

between the ends of a standard low resistance through which the current to be measured is passing. The wire is standardised with the Clark cell and then the low resistance R (say $\frac{1}{10}$ or $\frac{1}{100}$ ohm) is connected in series in the circuit in which the current is to be measured. The ends of this standard resistance are then treated as the terminals of a cell and joined up to the wire as shewn in Fig. 43. The balance is found and the volts determined. The current in R and therefore in the

circuit $= \dfrac{\text{E.M.F. across } R}{\text{resis. of } R}$, and since both are now known the

current strength is calculated.

Measurement of Resistance by Potentiometer. This is merely an elaboration of the fall of potential method and consists in measuring the E.M.F.s down a known and an unknown resistance through which equal currents are passing.

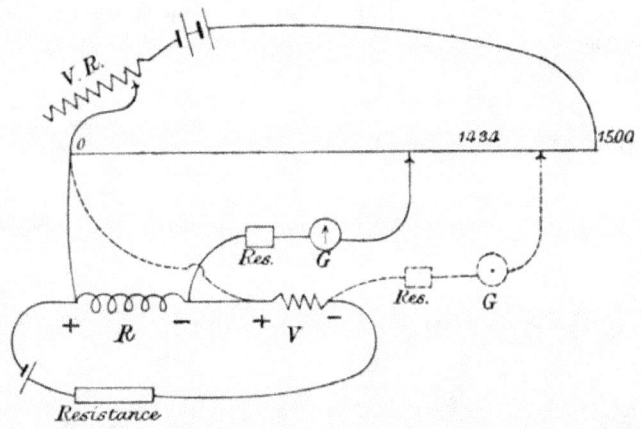

Fig. 44.

The wire is standardised, and then the unknown resistance r is connected in series with a standard resistance R, a cell and some resistance which may be varied. The E.M.F.s across R and r are then measured on the potentiometer when a constant current is flowing through them. Thus the unknown r can be determined. The connexions are shewn by Fig. 44.

The practical forms of potentiometer differ more or less in mechanical construction. They are all made to enable readings to be taken quickly and accurately ; and all made to give an equivalent to 1500 units of wire. However, there are in reality only 100 units of wire ; and 14 resistances each equal to the resistance of 100 units of the particular wire are arranged in series with it, and a switch is provided by means of which any number of these may be put in series with the wire on which the balance must be obtained. They are all self-contained, having a variable resistance and rheostat, and a number of pairs of terminals with a change switch so that a series of different measurements can be taken one after the other in a very short space of time. There is generally a micrometer screw arrangement for the fine adjustment of the jockey when a balance is nearly obtained. The E.M.F. of a Clark cell at different temperatures is also marked on one side of the scale.

CHAPTER V.

HEATING EFFECT AND ITS APPLICATIONS.

It has been pointed out that every circuit must offer resistance to the transmission of electricity and therefore energy must be expended in overcoming the resistance. The energy so expended cannot be destroyed but it is changed into Heat energy which may be useful or useless according to circumstances. When the energy of the current is needed for motive power then the Heat energy developed is useless and is consequently waste energy. The result is that the amount of work in the form required is less than it would have been had the whole of the energy of the current been transformed into that form.

The resistance of the bearings of an engine necessitates the absorption of part of the energy used to drive the engine. That portion is changed to heat energy—and the energy of motion obtained from the engine is less than the energy put into it by the amount of energy used to overcome the friction of the bearings.

This analogy must not be carried too far, for the precise nature in which resistance is offered to electricity cannot be stated. It is not known how electricity is transmitted through a circuit ; and it is impossible to compare resistance offered by "friction" to matter in motion, with resistance offered to electricity—the nature of which is not known. But the amount of energy obtained from a circuit in one form will always be less than the amount put into it, by the amount required to overcome the resistance of that circuit. This deficient amount will shew itself as Heat Energy.

Definitions of the Practical Units in terms of Work.
In order to appreciate the importance of this it is necessary
to consider the modes of measurement—or rather of definition
—of the units of current strength and E.M.F. It is shown
in the following pages that these units may be reduced to the
fundamental units of measurement—namely of length, mass,
and time.

It will be advisable, perhaps, to begin with these funda-
mentals and work up to the units of current and E.M.F. as
they are known.

Electricians work on the centimetre-gramme-second system,
generally known as the C.G.S. system. In this system the
centimetre is the unit of length, and a centimetre is the one-
hundredth part of a *metre* which is 39·37 inches. A centimetre is 0·3937 inch or an inch is 2·54 centimetres.

The unit of mass is the *gramme* and this is the mass of
1 cubic centimetre of water at its temperature of maximum
density which is 4° C. A gramme is $\frac{1}{453\cdot6}$ of a pound.

The *second* is the unit of time.

On the British system of units—unfortunately used by
mechanical engineers—the *foot* is the unit of length, the pound
the unit of mass and the minute the unit of time.

The C.G.S. system can be converted into a F.P.M. system
since the relationship of each to each is known.

Force produces or tends to produce motion or alteration
of motion of the body on which it is acting. A force acting
continuously on a given mass produces acceleration of that
mass ; and the acceleration produced is directly proportional
to the force and inversely proportional to the mass. A *unit
of force* is therefore defined as that force which acting continuously produces on a unit mass of matter, free to move,
unit acceleration. On the C.G.S. system the unit of force is
that force which, acting continuously on a mass of 1 gramme
free to move produces an acceleration of 1 centimetre per sec.,
per sec. This force is equivalent to $\frac{1}{981}$ of a gramme weight
and is called a *Dyne*.

Acceleration is change of velocity per unit of time, and

on the c.g.s. system the unit of velocity must be 1 centimetre per sec., and the unit of acceleration therefore (1 centimetre per sec.) per sec.

Work is done by a force moving its point of application in its own direction continuously against a resistance and producing motion of the body acted upon. The work is measured by the product of the Force and the distance through which it was uniformly acting. A *unit of work* is done when the unit of force is overcome through the unit of length.

Therefore on the c.g.s. system when a force of *one dyne* acts continuously through a distance of *one centimetre*, there is a unit of work done. This unit might be called a *dyne-centimetre* of work. It is called an *erg*. This unit is very small and too small for practical use. A multiple of an erg is therefore used in practice and this is = 10,000,000 ergs, and is called a JOULE.

A given amount of work may be done quickly or slowly. There is thus a *rate of working*. Rate of working is known as POWER, and the unit of power is the rate of working when one unit of work is done every second.

On the c.g.s. system this unit of power is 1 *erg per second*, with the fundamental units. With the joule as the unit of work the unit of power is 1 *joule per sec*. And *one joule per sec.* is called a WATT.

On the British system the units of work and power are the *foot-pound* and the *foot-pound per second* respectively. But a higher unit of power is generally used by engineers—called the *horse power*. This is the rate of working when 550 *foot-pounds of work are done per second*; or when 33000 *foot-pounds of work are done per minute*.

The student should compare the watt with a horse-power. He will find that 746 watts are equivalent to 1 horse-power.

The c.g.s. system may be summed up thus :

    Unit of Length   = 1 centimetre.
    Unit of Mass   = 1 gramme.

Unit of Time $= 1$ second.

Unit of Force $= \frac{1}{981}$ gramme weight $= 1$ dyne.

Unit of Work $= \frac{1}{981}$ gramme centimetre

                           $= 1$ dyne centimetre $= 1$ erg.

Unit of Power $= 1$ erg per sec.

Practical Unit of Work $= 10,000,000$ ergs $= 1$ joule.

Practical Unit of Power $= 1$ joule per second $= 1$ watt.

One Horse-Power $= 746$ watts.

In order to raise the temperature of any substance, work must be done. The *unit of heat energy* is the heat required to raise the temperature of a unit mass of water through a degree of temperature.

On the c.g.s. system a unit of heat energy is the quantity required to raise the temperature of 1 gramme of water through one degree centigrade.

Now mechanical work can be converted into Heat energy and Dr Joule found by experiment that 42,000,000 ergs of work converted into Heat energy produce one gramme-centigrade unit.

The work,—42,000,000 ergs or 4·2 joules—is known as the *mechanical equivalent of heat energy*, and since one unit of heat energy is equivalent to 4·2 joules of work, therefore H units of heat will be equivalent to $H \times 4\cdot2$ joules.

The unit of heat used by mechanical engineers is the heat energy necessary to raise the temperature of 1 pound of water through 1 degree Fahrenheit of Temperature. This is known as the British Thermal Unit, and its mechanical equivalent is 778 foot-lbs. of work. The British Thermal Unit is equivalent to 252 gramme-centigrade units of heat.

Work and Power in an electric circuit. In Chapter I on page 8, a volt is defined as the difference of potential between two points when in order to bring a coulomb of electricity from one point to the other it was necessary to do 10,000,000 ergs of work.

It might also be defined as the difference of potential

between two points when one coulomb of electricity flowing from one to the other gave out 10,000,000 ergs (or 1 joule) of work.

If 10 coulombs were moved then 10 times the amount of work would be done. If the difference of potential were increased then the work done would increase in proportion. Thus the work done varies as the difference in potential and as the quantity moved.

Thus E (volts) $\times Q$ (coulombs) $= U$ joules of work.

Now if this work be done in one sec. then the rate of working or the *power* would be joules per second—namely *watts*. But if one coulomb passes any cross section of a circuit in one second the strength of the current is said to be one ampere. And if a current of C amperes is flowing then the quantity per second being transmitted will be numerically equal to C coulombs.

Therefore if E again represents the difference of potential

E volts $\times C$ amperes $= E$ volts $\times C$ coulombs per sec.,

$\therefore E$ volts $\times C$ amperes $=$ joules per sec. $= watts$.

That is to say work is being done at the rate of EC joules per second. Therefore the power is EC watts.

Again,

In 1 second the work done is EC joules.

$\therefore$ In t secs. the work done is $E \times C \times t$ joules.

Hence in any circuit, the E.M.F. of which is E and the current in which is C, the work done per sec. is EC joules and the total work for any time t seconds is ECt joules.

Or, since $E = C \times R$ where R is the resistance between the ends of the circuit whose E.M.F. is E and in which the current is C, it follows that

$$E \times C \times t = C \times R \times C \times t$$
$$= C^2 \times R \times t.$$

Or again since $C = \dfrac{E}{R}$ it follows that

$$E \times C \times t = E \times \frac{E}{R} \times t = \frac{E^2 \times t}{R}.$$

Hence there are three expressions for the work done in t secs. by an E.M.F. of E volts maintaining a current of C amperes against a resistance of R ohms. The work done will be

$$E \times C \times t = C^2 \times R \times t = \frac{E^2 t}{R} \text{ joules.}$$

Determination of the relationship between the heat developed and the electrical work expended. This may be proved experimentally by measuring the heat energy developed in a resistance in a given time.

A convenient form of apparatus is shewn on Fig. 45.

The containing vessel is a copper calorimeter, consisting of two concentric copper cylinders, one suspended inside the

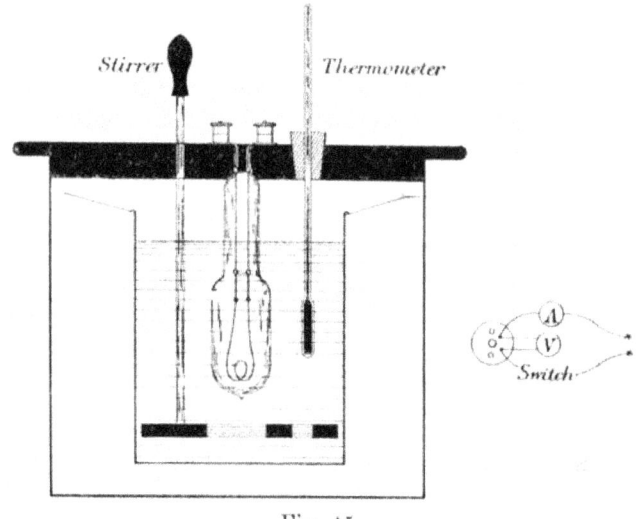

Fig. 45.

other by silk threads. The vessels are highly polished so that a minimum loss of heat by radiation takes place.

The inner vessel is weighed, and partially filled with

water.   It is weighed again and the weight of water thus obtained.

An incandescent lamp—constructed as shewn in the figure —is hung from a wooden lid, which fits over the outer vessel. This lamp is immersed below the surface of the water, and is connected to two terminals on the lid.   A thermometer is passed through a hole in the lid as shewn and the apparatus is completed with a stirrer.

To perform the experiment, the weight of the water and the equivalent weight in water of the calorimeter are found. (The water equivalent of the vessel is its mass × its specific heat.   In the case of copper, whose specific heat is 0·1, the equivalent weight in water of the calorimeter will be one-tenth of its weight.)   The apparatus is adjusted and the temperature of the water is taken.

The lamp is connected up to the terminals of a 100 volt circuit (if the lamp is designed for such a pressure) as shewn, an ammeter being connected in series and a voltmeter being connected to the lamp terminals.

At a known time the current is switched on and is kept on for, say, 10 minutes exactly.   During this time the water must be stirred continuously, and periodic readings of current and E.M.F. should be taken—say every minute.

At the end of the time the current is switched off and the temperature noted.

Now the heat developed will be equal to the mass of the water × the rise of temperature produced.   Let this $= H$ units. It has been shewn that $H \times 4\cdot2 = $ joules of work.

The work done by the current during the 10 mins. or 600 secs. will be $E \times C \times t$ joules.

The two quantities should be equal.

The following results were obtained by experiment :

Mass of calorimeter (inner vessel only)
    empty $= 133$ grammes
Mass of calorimeter + water $= 452\cdot6$ „
Mass of water only $= 319\cdot6$ „

Mass of water + equivalent water
    mass of calorimeter $= 319 \cdot 6 + 13 \cdot 3$
                                $= 332 \cdot 9 \text{ grammes} = 333.$

Initial temperature of water $= 16 \cdot 4° \text{ C.}$

Final temperature of water after 600
    seconds $= 36° \text{ C.}$

∴ Rise in temperature $= 19 \cdot 6° \text{ C.}$

  ∴ Heat given to the water $=$ corrected mass × rise in
                                             temperature
                        $= 333 \cdot 0 \times 19 \cdot 6$
                        $= 6525 \text{ units.}$

  ∴ the work equivalent $= 4 \cdot 2 \times 6525$
                        $= \mathbf{27,405} \textit{ joules of work.}$

Average current strength during the 600 secs. $= 0 \cdot 46$ amps.
Average E.M.F.   ,,   ,, $= 100$ volts.

∴ Work done by the current during the 600 secs. $= E \times C \times t$
                    $= 100 \times 0 \cdot 46 \times 600$
                    $= \mathbf{27,600} \textit{ joules of work.}$

It is seen that the results are very nearly equal. Hence the law known as Joule's law is proved, the heating effect of a current being proportional to square of the current strength, the resistance and the time.

Thus in any series circuit the heat developed in its various parts will be proportional to the resistances of those parts. The heat in each part will also vary in proportion to the square of the current strength, and the seriousness of increasing the strength of the current in any circuit above that for which it is designed will be apparent at once.

The fact that the heating effect varies as the *square* of the current strength and the resistance is readily accounted for. In order to double the current through a constant resistance it is necessary to double the E.M.F. Thus the rate of working will be increased fourfold since power is the product of E and C. If the current be increased three times, the E.M.F. will have to be increased three times, hence the power, $E \times C$, will be ninefold. And obviously the heat produced must vary as the power expended. The power expended varies as $C^2 R$, since $C^2 R = EC$, and if R be constant the power varies as C^2.

Thus it is more correct to say that the heat energy produced is directly proportional to the work done. This work $= E \times C \times t = C^2 \times R \times t$. Hence if R and t be constant the work varies as C^2. If C and t be constant the work varies as E, but it must be remembered that if E be doubled then to keep C the same R must be doubled. And whenever a statement is made that the heating effect varies as some one quantity it should always be completed by adding " when the other quantities (which should be named) are constant."

Board of Trade Unit of Electrical Energy. Electrical energy is bought and sold. It is the energy—the work which can be got out of the electricity—which is the thing of commerce and not the electricity itself. The energy of a supply for a given time is $E \times C \times t$ joules, when t is expressed in seconds.

The Board of Trade unit of electrical energy used for commercial purposes is one thousand watts for an hour. One thousand watts is called a *kilo-watt*, and the Board of Trade unit is spoken of as a *kilo-watt hour*.

Thus a current of 10 amperes at an E.M.F. of 100 volts supplied for 1 hour would yield a definite amount of work, called a kilowatt-hour and equal to 1000×3600 joules of work. The same amount of energy would be given out by a current of 100 amps. at an E.M.F. of 10 volts during one hour ; or by a current of 20 amps. at an E.M.F. of 100 volts for half an hour.

Hence Board of Trade commercial units of energy $= \dfrac{\text{amperes} \times \text{volts} \times \text{hours}}{1000}$, and electrical energy is bought and sold at so much per one of these units.

A dynamo is working at a full load. Its E.M.F. is 230 volts

and the current generated is 220 amps.   The power of the
dynamo is therefore $230 \times 220 = \mathbf{50{,}600}$ *watts*, or 50·6 kilowatts.

If this be run for five hours, then the total work done will
be 253,000 watt hours or 253 kilowatt hours.   That is 253
Board of Trade units of work.

If this is being sold at 4*d*. a unit the cost will be 1012
pence or £4. 4*s*. 4*d*.

Incandescent Electric Lamps.

When the heat generated by a current in a given resistance
is sufficiently increased the resistance may reach the temperature of red heat, or if it be of suitable material, it may be
made white hot.   When a wire is raised to the temperature of
incandescence a portion of the energy is changed into light
energy.   This is the principle of the commonest scheme of
electric lighting.   A filament is adjusted to offer such a
resistance that when a given E.M.F. is applied to its ends a
current is generated of sufficient strength to raise that filament
to the temperature at which it gives out light.

A portion of the electrical energy is thus changed to light
energy, but this portion is only about 5 per cent. of the whole,
the remainder being changed to heat energy !

The number of substances available for this purpose is very
limited.   Most conductors of electricity would melt or volati-
lise at a temperature lower than which would be required for
white heat.   Platinum, iridium and some alloys of these are the
only substances available, for these have a high melting point
and do not burn away in the air.   Unfortunately these are
expensive, and their melting point is very little above the
temperature of white heat, and if the applied E.M.F. increased
slightly and so increased the current the melting point might
easily be reached and the wire would fuse.

The substance commonly used is carbon.   Carbon has a high
specific resistance and a very high melting point, but it burns
away if heated up in the air.   This, however, can be prevented

by enclosing the carbon filament in a glass globe and exhaust-
ing the air therefrom. There will be no oxygen in the bulb,
and consequently the filament will be prevented from burning
when it is heated. That is the course adopted in practice.

In the design of a lamp it is desirable that the filament
should offer a high resistance, for the heating effect depends
upon C^2R. The current should be small since one does not
wish to heat the wires connected up to the lamp, and one does
not want the expense of putting in larger wires than are
absolutely necessary. Hence the resistance of the filament
should be such that when a given E.M.F. is applied to its ends,
the current generated in that resistance should just raise its
temperature to that of incandescence.

Manufacture of Carbon Filament Lamps. The
carbon filament is made by dissolving filter-paper or cotton-

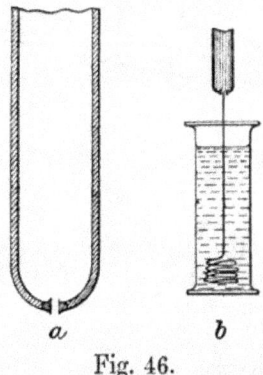

a   b

Fig. 46.

wool in zinc chloride to make a jelly-like substance of the
consistency of treacle. This substance is called *cellulose*.

The cellulose is next formed into a long string. This is
done by forcing it through a "squirt." This consists of a
glass tube through which the cellulose is forced under a
pressure of about 20 inches of mercury. At the end of the

tube is a "die"; and this die is a ruby through which a hole is pierced of the diameter required for the particular filament which is being made.   This is shewn in section by Fig. 46 a.

The cellulose thread thus formed passes down into a vessel of alcohol in which it coils itself up, as is shewn in Fig. 46 b.   The alcohol washes the cellulose and hardens it somewhat.   It is allowed to remain in the alcohol for about 28 hours, when it has assumed the appearance of catgut and will be of similar toughness.   It is then washed thoroughly in water.

The next stage consists in carbonising the cellulose.   This is done by fastening lengths on to a block of carbon as shewn in Fig. 47.

This carbon block is formed so that the filament is shaped as required and is of the length necessary for the lamp.   In

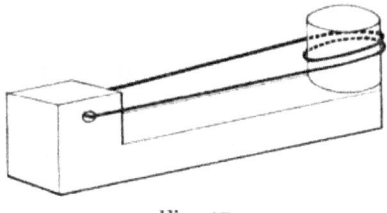

Fig. 47.

the case illustrated the filament is looped once round the upright carbon cylinder at the end of the block.   The ends are stuck on the other end of the block with some carbonaceous cement.

It should here be mentioned that in the process of carbonising the cellulose shrinks considerably.   Hence some precautions have to be taken and due allowances made.

The carbon blocks with their cellulose threads are then packed in an air-tight crucible, with powdered charcoal filling up the spaces, and the whole is slowly raised to a white heat in a gas furnace.   They are kept at this temperature for about 9 hours.   This high temperature reduces the cellulose to carbon ; renders the carbon hard although brittle ; increases its conductivity ; and expels occluded gases from its pores.

Pieces of platinum are then fixed to the ends of the formed carbon filament. These platinums are to be fused through the glass bulb ultimately and serve as conductors for the current. The fixing is done by flattening out the ends of the platinum wires and looping the flattened ends round the filament. A little carbonaceous cement is also used, and the joints are made better in many cases by short-circuiting the filament and dipping the whole in a hydro-carbon oil and passing a current through the platinums. This heats up the junctions and carbon is deposited upon them. The operation is attended with a little danger and the oil tank is generally provided with an air-tight lid which can be shut down immediately on any sign of combustion.

The filament is then "flashed." The flashing process consists in passing a current, at double the E.M.F. to be subsequently used, through the filament which is placed in an atmosphere of some hydro-carbon—generally coal gas.

The filament becomes hot and the coal gas is decomposed, carbon being deposited on the filament. If the filament is uneven then the thinner portions will become hotter than the thicker, and on these portions carbon will be deposited at a greater rate. After a little while the filament will thus become uniform.

Flashing is used rather as a test of, than as a means of producing, uniformity now-a-days, for the method of making the filament does not produce the irregularities which were obtained when the filament was made by parchmentising cotton.

Flashing is also used to make the final adjustment of the resistance of the filament. The latter is made one arm of a Wheatstone bridge and flashed until a balance is obtained at the proper resistance.

The filament is then fixed in the glass bulb, the pieces of platinum being fused through at the top of the bulb, as shewn in Fig. 48.

The remaining open end of the bulb is then connected to
a mercury vacuum pump and the air is exhausted gradually.
While this is going on the bulb is heated with a Bunsen flame
to expel any air film on the glass.   When the exhaustion

Fig. 48.

has reached a high stage the filament is heated also, by the
passage of a current.   This rids the carbon of occluded gases,
hardens it and further increases its conductivity.

When exhaustion has reached a maximum the bulb is
sealed off with a fine blow-pipe flame.   It only remains to fix
the brass collar for connexion purposes and the lamp is
completed.   The collar holds two brass segments, insulated
from each other and from the collar itself by means of plaster
of Paris cement.   Each segment is connected to one of the
platinum wires making contact with the filament inside the
lamp.

When completed the average carbon filament lamp requires
about 3·5 watts for every candle power of illumination it
yields, or it will yield 0·286 candle power per watt.

The vacuum of a completed lamp is tested usually by
holding the bulb in the hand and bringing up one terminal
of the lamp to a terminal of an induction coil.   If a glow is
produced within the lamp the vacuum is not sufficiently good.

Osmium and Tantalum Lamps.   There are two forms
of incandescent lamp which have been put upon the market

8—2

lately having metallic filaments. These are the "Osmium" and "Tantalum" lamps respectively, and it is claimed for them that they absorb less energy for every candle power they yield than is absorbed by the carbon lamp.

The lamps are similar in general appearance to the carbon filament lamp ; but their filaments are much longer, being of lower specific resistance. In both cases the air is extracted from the bulb.

The osmium lamp has a filament of the metal osmium. The light produced is much whiter in quality than that of the carbon lamp, and no blackening of the bulb is produced. They yield 0·6 candle power for every watt absorbed but their average life is only one-half that of the ordinary lamp, and their first cost is considerably greater at present. Notwithstanding this there are cases in which it would be more economical to use one rather than the other and examples are shewn on page 125. The osmium lamp is only made for low voltages at present and this generally necessitates that groups of 2 or 3 in series must be used. This is disadvantageous in that it involves the use of special fittings ; and moreover if one of the lamps should burn out the other members of the group will be cut off from the current supply.

The tantalum lamp claims similar advantages to the osmium, its candle power per watt being the same, and the quality of its illumination being of the whiter order. It is made for voltages up to 110 at present. Its first cost is between that of the carbon and the osmium lamp, and its average life is given as 500 hours.

The Nernst Lamp. The Nernst lamp is the solitary example of an incandescent filament which is not enclosed in a vacuous space.

The filament of the Nernst lamp is composed of a mixture of such metallic oxides as magnesia, yttria, zirconia, ceria or thoria. This filament does not *burn* when raised to the temperature of white heat in air.

However it is an insulator at ordinary temperatures, but

when raised to a high temperature it becomes a good conductor and a good luminant. It follows therefore that when a finite E.M.F. is applied to the cold filament, its resistance being high, there will not pass a sufficiently great current to produce any appreciable heating. External means have to be adopted then for heating the filament, and when the necessary temperature

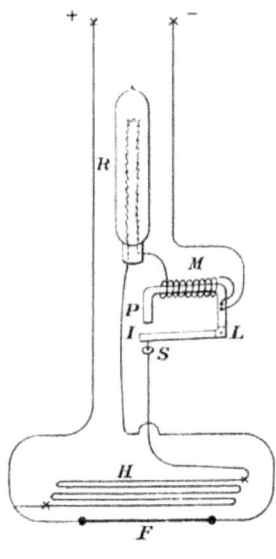

Fig. 49.

has been attained a current may be passed through the filament which will become brilliantly luminous.

The filament might be initially heated with a spirit flame ; but in the Nernst lamps now in general use the heating is effected by means of an iron wire resistance which is covered with a thin coating of the oxides of which the filament is made.

Fig. 49 is a purely diagrammatic illustration of a typical Nernst lamp. The current can pass from the positive terminal through the filament F and also can branch through the heater H. The other end of the filament is connected to a lamp resistance R (which consists of a length of fine iron wire in a vacuous bulb), and from there through the coil of a small electro-magnet M. The other end of the coil is connected to the iron core. From this point there is connexion either to the negative terminal or through the core to the stud S and thence to the other end of the heater.

A piece of iron IL is hinged at L and makes contact with the stud S.

When the switch is put on no current will pass through the filament because of its high resistance. But a current will pass through the heater H. This will raise the temperature of the filament immediately beneath it and after a little time a current will pass through the filament and produce incandescence. But when this occurs the current must also pass through the resistance R and magnet coil M. This will pull up the iron IL to the pole P, and thus break contact at the stud S, when the heater will be cut out of the circuit.

The efficiency of the lamp is 2—2·5 watts per C.P. The first cost of a lamp is somewhat great, but the different parts can be renewed separately. The lamp does not light up for about a minute after switching on.

The resistance R is a steadying resistance, and being of iron it increases with the temperature ; on the other hand the filament resistance decreases. Thus if the E.M.F. should be increased, the lamp resistance would increase and balance off the decrease in filament resistance, and preserve an approximately steady current. If the lamp resistance was not used it can be seen that an increase in applied E.M.F. would produce a great increase in current due to the large decrease in the filament resistance.

Efficiency of Incandescent Lamps. The efficiency of an incandescent lamp is measured by the candle power per watt, but more generally expressed in terms of watts per candle power.

The average efficiency of a carbon filament lamp is 3·5

watts per candle power for those whose candle power ranges from 8 to 50; and about 2·5 for those above 100 c.p.

The candle power of a lamp is measured by comparing the illumination of the lamp with that of a standard of candle power. The process is known as *photometry*, the apparatus used constituting a *photometer*.

The principle of the measurement is briefly as follows. A screen is placed between the two sources of light and adjusted with respect to distance so that both sides of it are equally illuminated—one by the standard and the other by the lamp whose c.p. is to be determined.

The illumination per unit area depends upon the intensity or candle power of the source of light and upon the distance of the illuminated area from that source. It varies directly as the candle power (generally written c.p.) and inversely as the square of the distance.

Thus illumination per unit area $\propto \dfrac{\text{c.p.}}{\text{distance}^2}$.

In the measurement, the screen is so adjusted that the intensity of illumination on each side is the same.

Hence it follows that

$$\frac{\text{c.p. of the lamp}}{(\text{distance between lamp and screen})^2} = \frac{\text{c.p. of standard}}{(\text{distance between standard and screen})^2}.$$

$$\therefore \text{c.p. of lamp} = \text{c.p. of standard} \times \frac{(\text{distance between lamp and screen})^2}{(\text{distance between standard and screen})^2}.$$

Photometers. There are many different forms of photometers, but the underlying principles of each are the same. In accurate measurements the lamp under test is mounted so that it is capable of rotation about a vertical axis, and it is rotated at a fairly high speed during the process of adjustment. The standard of light may be a standard candle (made of spermaceti—weighing $\frac{1}{6}$ lb.—and burning at the rate of 120 grains per hour) whose candle power is unity; or it may be the Pentane standard; or it may be an incandescent lamp specially made and standardised previously.

The important part of the photometer is the " photometer

head"—mentioned above as the screen. In Bunsen's photometer, this screen consists of a small piece of paper having a
uniform "grease spot" at its centre. This spot is translucent, and when light falls upon the screen the spot appears
darker on the illuminated side and brighter on the remote
side. Now when the disc is illuminated on both sides the

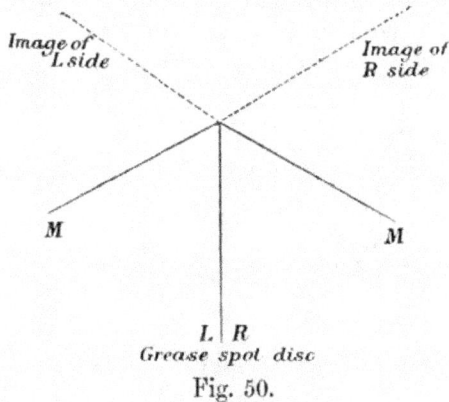

Fig. 50.

grease spot will apparently disappear when the intensity of
illumination is the same on each side. And the position of
the disc is adjusted until this result is attained. An arrangement of mirrors will enable an observer to see both sides of
the disc at once, as is shewn by Fig. 50, which is a diagrammatic
plan.

A very sensitive and largely used photometer head is that
known as the *Lummer-Brodhun*, which is illustrated by Fig. 51.
A white screen S is placed in a metal box so that it can be
illuminated on both sides by sources of light at A and B.
Two right-angled reflecting prisms P_1 and P_2 reflect the light
from the opposite sides of the screen as shewn to a compound
prism $P_3 P_4$. This consists of a cube of glass which has been
cut into two parts diagonally, forming two right-angled prisms.
The prism P_4 has its hypotenuse surface rounded off at the
corners as shewn, and the remaining part is optically worked

to make optical contact with the hypotenuse surface of
P_3. The path of the light is shewn from both sides of the
screen, and the observer looking through the telescope T will
see a circular patch of light surrounded by another zone of
different intensity. The circular central patch will be due to
the light transmitted through the centre of the compound
prism after reflexion from P_1. The outer light is that totally
reflected from the hypotenuse surface of P_3, after reflexion
from P_2. The positions of A and B are adjusted until the
field of vision becomes quite uniform.

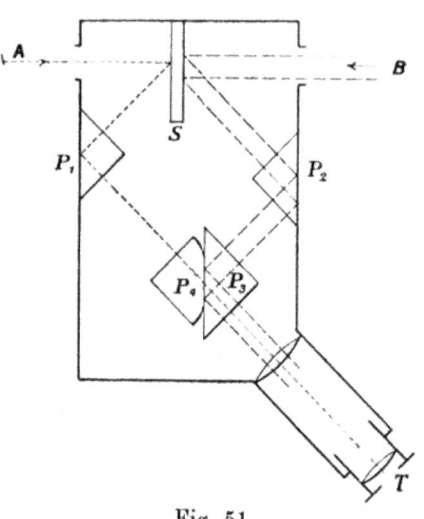

Fig. 51.

Another form is that known as the *Flicker* Photometer,
the principle of which is illustrated by Fig. 52.

S is a fixed white screen illuminated by one source A.
F is a rotating "fan" in the form of a maltese cross as
shewn by the figure marked F. This cross is made of white
pasteboard like the screen S and is illuminated by the second
source B. Its open sectors are equal to the closed sectors.

The screen and fan are placed as shewn and the observer

looks through the telescope at T. If the fan and the screen are equally illuminated a uniform and steady field of view will present itself, but if either S or F be brighter than the other the field of view will present a *flickering* appearance—hence the name of the photometer. Adjustment of A and B is therefore made until there is no flicker observed.

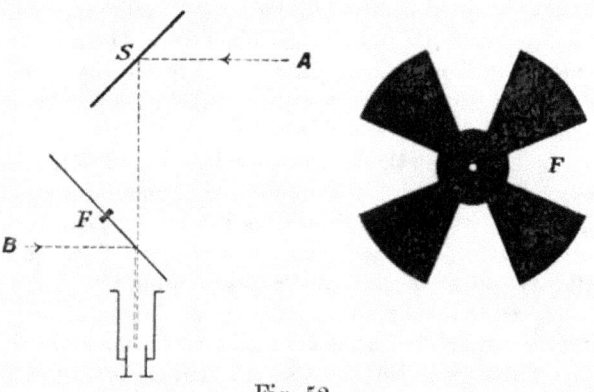

Fig. 52.

In measuring the c.p. of different types of lamps some difficulty is found in making accurate adjustments due to the different colours of the source and the standard. When this occurs adjustment should be made by the observer looking through a red glass, and secondly through a blue glass. The mean c.p. for the two adjustments will give the required measurement.

Determination of Efficiency. In the determination of the efficiency it is necessary that a constant E.M.F. and therefore a constant current be maintained. For this reason the source of E.M.F. usually consists of a battery of secondary cells which will yield the potential difference for which the lamp was designed. When the photometer is adjusted the E.M.F. across the lamp terminals and the strength of the current in the lamp are determined simultaneously. The product of these gives the total watts consumed.   The watts per c.p. are then calculated.

The expression "the efficiency of a lamp" hardly corresponds to the similar expression used in connexion with an engine, or a dynamo, or a battery.   In the latter cases the word efficiency is used to indicate the ratio of work given out to work put in—or useful work to total work.   But in the case of the lamp it is used to indicate usually the ratio between the power put in and the candle power obtained ; or more properly, though less generally, between the candle power obtained and the power put in.   Since a unit of candle power has not yet been defined in terms of a quantity of work per unit time, it is seen that the number denoting the efficiency of a lamp cannot have any absolute significance concerning that efficiency ; it can only be used as a means of comparison with other lamps.   The comparative numbers will be quite proportional to their relative efficiencies however, but it must be remembered that they give no idea of absolute efficiency—that it to say they give no idea of how much energy in the form of light is given out by the expenditure of a unit of electrical energy.   Or, putting the matter on a commercial basis, although the cost of electrical energy put in is readily determined, the so-called "efficiency" does not enable one to calculate how much of that cost is given back in the form of light.   Is one getting one's money's worth ?

The Proportion of the Electrical Energy given out as Heat.   As a matter of hard fact the production of light energy, *via* the medium of heat energy, from electrical energy is a somewhat inefficient performance.   Experiment goes to prove that of the electrical energy put into the lamp only about 5 to 7 °/₀ appears in the form of light—the remaining 95 or 93 °/₀ appearing as heat energy.   That is to say, for every sovereign expended only one shilling's worth of light is produced.

It should be added hastily that the ordinary gas jet and oil flame are no more efficient in this respect.

The fact that some 95 °/₀ of the total energy put into a lamp appears as heat is often questioned on first statement. There is a popular impression concerning the "coolness" of the electric light which dies hard. It has probably arisen through the fact that the incandescent lamp does not absorb any of the oxygen of a room and consequently does not contribute to that stuffiness which is so often mistaken for warmth. But the use of electric radiators—merely large lamps—for warming purposes should serve to dispel the illusion.

The proportion of energy which is converted to the form of light may be determined experimentally in a manner similar to that described earlier in this chapter on page 108. The experiment is modified by having a *glass* calorimeter in place of one of metal. Its water equivalent is determined, and the experiment is carried through in exactly the same manner as described before; and the quantity of heat energy produced in a given time is determined.

The lamp is then rendered opaque with a covering of lamp-black varnish, and the experiment is repeated under the same conditions. The quantity of heat energy produced in the same time is thus determined.

Now it will be found that in the latter case there will be more heat energy than in the former case, although the E.M.F.s, currents, and times were identical.

It is concluded that in the latter case all the energy expended appears as heat, whilst in the former case a portion of it appears as light.

It will be found that

$$\frac{\text{Heat developed when lamp was transparent}}{\text{Heat developed when lamp was opaque}} = \frac{0\cdot95}{1}.$$

Therefore $\frac{0\cdot05}{1}$ of the total energy appeared as light, viz. 5 °/₀.

One might point out as some support to this, that a "frosted" bulb is always hotter to the touch than a clear bulb of an equal lamp. Coloured bulbs are also hotter. The bulbs of electrical radiators are obscured and it might be suggested that they would be more efficient as radiators if they were made opaque with black varnish. It should be remembered, however, that imagination enters largely into the sensations experienced—especially in England—and heat without an accompaniment of light is not considered desirable though possible.

Examples on cost of running Incandescent Lamps. In considering what kind of incandescent lamp it would be desirable to use in a lighting installation, there are four points to be taken into account. These are the watts per c.p. of the lamp, the duration of life, the first cost of the lamp, and the cost per Board of Trade unit of the electrical energy available.

Two sets of examples may be quoted and worked out, comparing two lamps, each of 25 c.p. A costs 1s. and lasts for 1000 hours with an efficiency of 3·5 watts per c.p.; B costs 7s. 6d., and lasts 500 hours with an efficiency of 1·5 watts per c.p.

Firstly, they may be compared for use at a place where the energy costs 6d. per b.o.t. unit.

Case I.   *Lamp A.*

$$\text{Total watts required} = 3{\cdot}5 \times 25 = 87{\cdot}5.$$
$$\text{Total watt-hours} = 87{\cdot}5 \times 1000 = 87500.$$

One b.o.t. unit = 1000 watt-hours.

$$\therefore \text{B.O.T. units required} = \frac{87500}{1000} = 87{\cdot}5.$$
$$\therefore \text{Cost of running for 1000 hours} = 87{\cdot}5 \times 6 = 525d.$$
$$\text{Initial cost} \quad . \quad . \quad . \quad . \quad . \quad = 12d.$$
$$\therefore \text{Total cost} \quad . \quad . \quad . \quad . \quad . \quad = 537d.$$

Lamp B. $\text{Total watts required} = 1{\cdot}5 \times 25 = 37{\cdot}5.$
$$\text{Total watt-hours} = 37{\cdot}5 \times 1000 = 37500$$

for the same length of time as A.

$$\therefore \text{B.O.T. units required} = \frac{37500}{1000} = 37{\cdot}5.$$
$$\therefore \text{Cost of running for 1000 hours} = 37{\cdot}5 \times 6 = 225d.$$

Two lamps would be required to run for 1000 hours, and thus the cost of the lamps would be $2 \times 7s.\ 6d. = 180d.$

$$\therefore \text{ Total cost} = 405d.$$

Hence in this case it would be more economical to use the type B, even though it costs much more originally and two lamps are needed for the period of 1000 hours.

CASE II. Same lamps, but the cost of energy supplied is $2d.$ per B.O.T. unit.

Lamp A.

Cost of running for 1000 hours $= 87\text{·}5 \times 2 = 175\text{·}0d.$
Cost of lamp . . . . . $= \ \ 12d.$
Total cost . . . . . $= 187d.$

Lamp B.

Cost of running for 1000 hours $= 37\text{·}5 \times 2 = \ \ 75\text{·}0d.$
Cost of two lamps . . . $= 180d.$
Total cost . . . . . $= 255d.$

Hence in this case it would be cheaper to use the lamp A.

Life of a Lamp. It is shewn that it does not always pay to run the high-efficiency (i.e. the low watts per C.P.) lamp, for much depends upon first cost and on the life of the lamp. Another point always worthy of consideration is the behaviour of the lamp during its life. Generally speaking, the efficiency of a lamp decreases with the number of hours it has burned. With a well-made lamp this increase in the watts required per C.P. is not very serious, but with some of the cheaper forms of so-called high-efficiency lamps the increase becomes so great that the *average* watts per C.P. during the life of the lamp is greater than that of a lamp which professedly is of lower efficiency at the outset. This is the case especially with some of the cheap German lamps which were introduced some little time ago. These professed to absorb but $2\text{·}0$ to $2\text{·}5$ watts per C.P. Their life was only 500 hours, however; and their average watts per C.P. during the life was in some cases higher than $3\text{·}5$. Clearly such lamps are expensive to run.

In constructing a lamp which shall absorb fewer watts per c.p., the filaments are made of higher resistance and very much thinner than the more normal form. They are run at a higher temperature also, and this tends to shorten their lives considerably.

An interesting matter in connexion with the running of lamps is the relationship between the E.M.F. at which they are run and their life. A 100 volt lamp is designed to give its rated c.p. at 100 volts. If the pressure applied be greater, then the c.p. also increases—but not economically, for the life of the lamp is considerably shortened, even though the watts per c.p. are decreased. If the lamp be run under normal pressure then the life is increased, but the watts per c.p. is increased very considerably, and it will be seen that if the E.M.F. be sufficiently reduced the current generated will not produce even a dull glow. At that stage there will be an absorption of power, but no return in the form of light.

Below is a table illustrating the relationship between the E.M.F., the c.p. and the life of a series of 100 volt 16 c.p. lamps.

E.M.F.	C.P.	Life in hours
96	10	2750
98	12	1645
99	14	1275
100	16	1000
101	18	785
103	20	475
105	22	285

When a lamp is some 600 hours old, its efficiency is very low, and generally speaking it would be more economical to replace such lamps with new ones at this time, for the cost per c.p. of running the old lamp is very considerable. It is at this stage too when a lamp could be economically *overrun*, for by overrunning the watts per c.p. would be reduced to a reasonable order even though the life would be shortened. The Americans use the term "smashing point" to denote that stage in a lamp's life (about 600 hours) when it would pay to break the lamp and put in a new one.

The relationship between efficiency, life and voltage is shewn by the following table.

Voltage	Efficiency	Life
98 °/₀	3·77 watts per c.p.	1200 hours.
100 °/₀	3·5 ,, ,, ,,	900 ,,
102 °/₀	3·27 ,, ,, ,,	700 ,,
104 °/₀	3·05 ,, ,, ,,	550 ,,
106 °/₀	2·87 ,, ,, ,,	440 ,,

The filament of a 16 c.p. lamp for 200 volts naturally differs from that of a similar 100 volt lamp. The c.p. produced depends upon the work done per second—upon the power expended electrically. This power is the product of the E.M.F. and the current. Consequently a 200 volt lamp will require only one half the current required for a 100 volt lamp of equal candle power.

Thus the resistance of the filament of a 200 c.p. lamp must be *four times* that of the equivalent 100 c.p. lamp. To get this the filament must be either longer or smaller in area of cross section, or both. The filament of a 16 c.p. 200 volt lamp might be equivalent to two 8 c.p. 100 volt filaments in series—and such a plan is frequently adopted.

High-pressure lamps are not as a general rule so durable as those of low pressure. A thin filament is more liable to give out before its 1000 hours than a thicker one. And there seems to be some difficulty in preventing "shorts" (short circuits) across the junctions between the carbon and the platinum. If a consumer has any choice of pressure (such is rare), he should take the low-tension.

CHAPTER VI.

ARC LAMPS.

WHEN the switch of a "live" circuit is broken a spark may be observed at the place where the contact breaks. Sometimes this spark is big, and under other conditions small. Switches are always constructed so that they "break" quickly, leaving a big air gap between the points of contact, for if they did not a continuous spark might be maintained which would fuse the metal work across which it was playing.

The maintenance of such a permanent spark forms the essential part of electric lighting by those lamps known as *arc lamps*.

Production of an Arc. The first permanent spark was produced by Sir Humphry Davy in 1810. He connected two sticks of wood charcoal to the terminals of a big series battery of some 2000 primary cells, the E.M.F. of which was between 2000 and 3000 volts. The charcoal sticks, arranged horizontally, were gradually brought nearer to one another endways, and when they were about $\frac{1}{40}$th of an inch apart a spark jumped across the intervening space. The spark was not momentary however — it apparently developed into a continuous and intensely luminous flame. Moreover when the charcoal sticks were moved further apart this flame continued to play until a space of over an inch separated the ends of the carbons.

The flame naturally formed an arch—hence the name *arc* which is given to this source of light.

It was found that once the arc had gone out it could not be restarted unless the carbons were brought close together again.

The temperature of this arc was found to be very high, higher indeed than that produced by any other known source, for such substances as platinum, diamond and carbon could be melted and even volatilised in it. The temperature of an arc has not been capable of determination up to the present time, but it has been estimated at about 3500° C.

The modern arc is not an exact reproduction of Davy's arc, for it was found that an arc could be maintained by a much smaller E.M.F. than 2000 volts. But although it can be maintained it cannot start itself unless the E.M.F. be of a fairly high order. With two pieces of carbon between which there is an E.M.F. of 45 volts a good arc may be produced. On bringing the ends together gradually no spark will be found to jump across even the smallest visible air gap. Experiment goes to prove that some 27,500 volts are needed to cause a spark to jump across a centimetre air gap. The air has an infinite electrical resistance, but when it is subjected to the mechanical strains produced by the electric field about 2 knobs, between which there is an enormous potential difference, it may break down due to some molecular change and allow a spark to jump across. But the air can withstand the strain produced by the electric field between two carbons having 45 volts across them, and no spark is produced on bringing them together.

But when touching the points of contact become hot. The air about them also becomes heated and this decreases both its resistance and its capacity for withstanding electric strain. When the carbons are separated a little a spark jumps across, probably aided by the altered electrical conditions of the air and by some extra E.M.F. of *self-induction*. This same spark produces a certain disintegration of the carbon, minute particles travelling across the space. These particles further reduce the electrical resistance of the space and become hot to the temperature of incandescence. It is now found that a steady current can be maintained across the carbons, that a brilliant light is produced, and that the carbon is being melted and volatilised by this "continuous flame." But the space between the carbons cannot be regarded as an air space. It is impregnated with carbon particles and with carbon vapour, and under such conditions it is possible to separate the carbons some quarter of an inch with an E.M.F. of 45 volts across them.

In the utilisation of this phenomenon for electric lighting the two carbons are generally placed vertically, one above the other. The appearance of the carbons and of the arc itself depends upon the nature of the E.M.F. supplied. This E.M.F. may be continuous and steady (always acting in the same direction), or it may be alternating and varying, that is to say changing its direction some 50 or 100 times per second.

If a continuous E.M.F. be employed it is found that positive carbon becomes very much hotter than the negative, and consequently burns away at a greater rate.

Formation of Continuous Current Arc. With continuous E.M.F. the positive carbon is usually placed above the negative, and its area of cross section is about twice that of

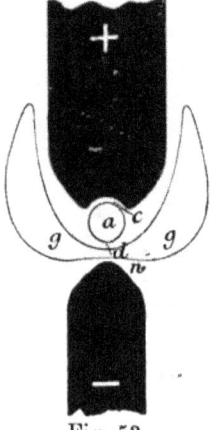

Fig. 53.

9—2

the negative since the rate of burning away is twice as great. Under these circumstances the carbons assume the formation illustrated by Fig. 53. The positive carbon resembles an inverted volcano on a small scale, the hollow portion being called the crater of the arc. This crater c becomes brilliantly incandescent and in the ordinary arc about 85 $°/_∘$ of the total light is emitted from the crater.

a represents the arc proper ; d is a dark space, and gg is a "flame" of incandescent particles of carbon carried upward by the convection currents of hot air. This "flame" is known as the *golden aureole*. n is the negative carbon, which does not become so brilliantly incandescent as the crater. It emits about 10 $°/_∘$ of the total light.

If the image of an arc be projected upon a screen by means of a lens the formation of the arc can be seen very clearly.

Properties of an Arc. An arc may be blown out as a candle is blown out ; and a high tension arc (say 2000 volts) may be "lighted" if the carbons are sufficiently near to one another. This latter process simply consists in warming the air between the carbons—or rather in substituting a number of incandescent particles (the flame) for the air. The high tension may then cause the arc to strike.

A naked arc may be burned under water, and at the same time the water will be decomposed electrolytically.

Another interesting point about an arc is that it behaves like a flexible conductor conveying a current. It has a magnetic field about it, and when a magnet pole is brought up it is deflected to the one side or the other according to the polarity of the pole presented. If the pole be brought near enough, then the arc may be deflected so much that it goes out.

If a pole of a small electro-magnet, through the coils of which an alternating current is passing, be brought up then the arc spreads out on both sides—hissing the while—and soon goes out as the alternating magnet is brought nearer.

The same effect is produced by bringing up a permanent magnet pole to an arc maintained by an alternating E.M.F.

E.M.F. of Maintenance of an Arc. It has been found that an E.M.F. of 44 volts is required to maintain an arc between *carbons*, although with other electrodes the E.M.F. required may be greater or less.

The distribution of potential along an arc has been determined by connecting one terminal of a voltmeter to the end of the negative carbon and having a little exploring pencil of carbon, Fig. 54, joined through a flexible wire to the other terminal. This pencil can be put at any point either on the

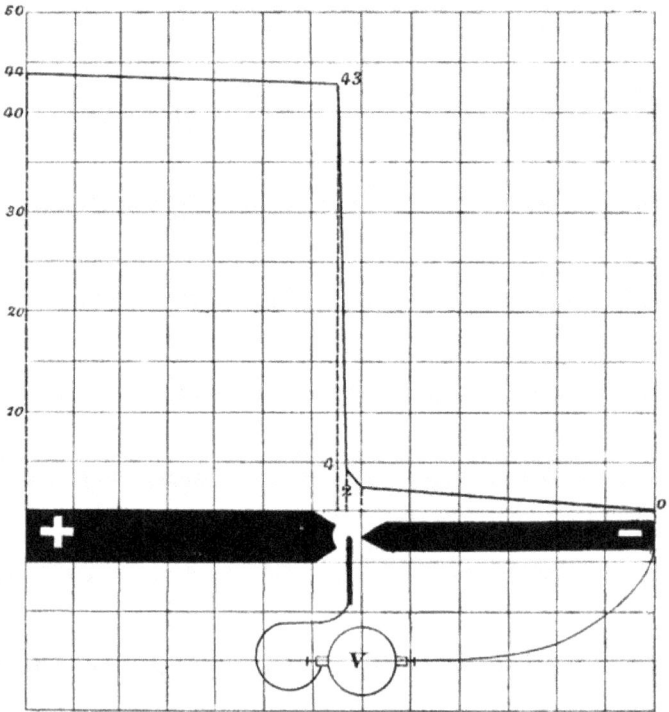

Fig. 54.

carbons or in the arc and the potential difference between that point and the end of the negative carbon measured. The results obtained are shewn by the diagram—the vertical heights representing the potential difference between each point and the end. It is seen that in this particular instance the potential difference across the negative carbon was 1·5 volts, and across the positive carbon 1 volt. The important point is the drop which is observed between the crater and a point almost immediately outside it. This is found to be 39 volts, the drop through the remainder of the arc being 2 volts in this case.

The figures given above are variable, depending upon the strength of the current, and the sizes of the carbons and the length of the arc, with the exception of the 39 volt drop between the crater and a point just outside it. This is found to be a constant—for carbon—and is independent of the size of the carbons, the strength of the current or the length of the arc. It is always absorbed—the remaining E.M.F. having to maintain the current through the carbons and the vapours between them. In the example given above 5 volts (viz. 44—39) is all that would be needed to maintain the current (i.e. 5 amperes) if the 39 volts did not happen to be absorbed.

It was thought that this absorption was due to a back E.M.F. like the back E.M.F. of polarisation, or of a running motor.

It is much more probable, however, that it has to do with the vaporising of the carbon. A certain amount of heat energy is necessary to melt, and to boil every gramme of carbon. The carbon is unmistakeably vaporised, but it has never been determined how much heat is necessary per unit mass. The probability is that 39 × current strength × time represents the energy required to vaporise a definite mass of carbon.

This view is supported by the fact that although the strength of the current may be increased to any extent the *temperature of the arc is not increased*, neither is the C.P. per unit area of the crater, although of course a greater area is heated to the temperature of incandescence.

Now it is well known that when a kettle of water is put upon the fire the temperature of the water does not increase at all when once boiling commences. The heat energy is then being used up in vaporising the water. Similarly it is assumed that the energy represented by $39 \times c \times t$ is being used up in vaporising the carbon. The remainder of the electrical energy is maintaining the arc. This point can only be definitely settled when the latent heat of vaporisation of carbon is determined, but it seems a much more probable explanation than the suggested back E.M.F.

An arc maintained by an alternating E.M.F. has a different formation to the continuous arc. Both carbons become equally hot and therefore equally luminous, and are made of equal cross sectional area. They get a conical formation with blunted ends and the various characteristics are difficult to define owing to the alternation.

The character of the light emitted from the *crater* is similar to sunlight, but has a slight *excess* of orange and green and a slight *deficiency* of blue. The fact remains, however, that the light from the whole arc appears to be purple. The space between the carbons (the arc proper) is impregnated with carbon vapour and carbon particles, which combination is called the arc *mist*. The light emitted from or through this mist is pronouncedly strong in violet. The probability is that the orange and green are absorbed in this mist after a series of internal reflexions, and since the light received from the crater must pass through this mist, the general character of the received light appears to be strong in violet and weak in orange and green. Mrs Aryton discusses this at some length in her book on *The Electric Arc*.

Automatic Feed Lamps. When an arc has been struck —that is when the carbons have been brought into contact and separated again—it burns steadily for about two minutes and then begins to flicker a little. The carbons are burning away and the length of the arc is increasing. Ultimately the arc would go out, but it can be maintained by " feeding " the

carbons together, and if this feeding be done at the same rate
as they burn away a perfectly steady arc may be maintained.
This is done automatically in those lamps used for general
lighting.

An automatic arc lamp must be so constructed that the
carbons are always brought in contact when no current is
passing. It must also separate the carbons—i.e. strike the
arc—to a certain workable distance when the current is passed,
and it must feed the carbons together gradually and practically
insensibly as they burn away.

It is not necessary to describe the various types of arc
lamps on the market. They are all very much alike in construction and the underlying principles are quite identical.

The Crompton lamp—illustrations of which are given—
serves to demonstrate the various questions at issue.

Crompton Arc Lamp: Open Type. The governing
forces for the working of this and other self-feeding arcs are

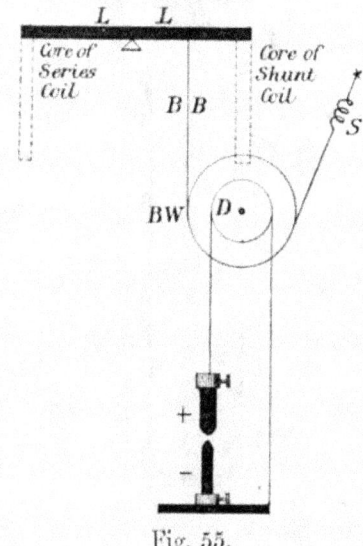

Fig. 55.

provided by two magnetic fields. One field is produced in a solenoid in series with the arc and the other in a solenoid which is really shunted across the two carbons. As the arc lengthens, due to the burning of the carbons, its resistance will increase. Therefore the current in the series solenoid will be decreased and at the same time the shunted current in the shunt solenoid will be increased.

These variations are utilised to regulate the feeding of the carbons. The series coil has a small resistance, whilst the shunt coil has a resistance of 200—300 ohms.

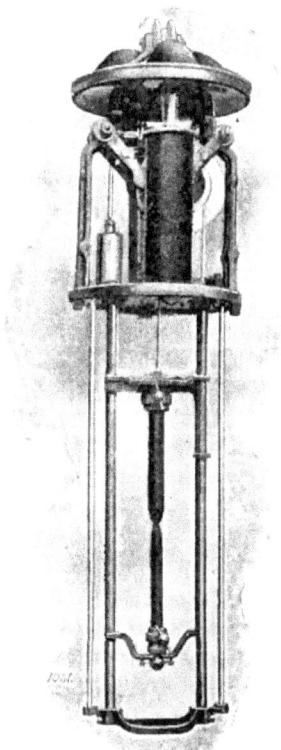

Fig. 56.

Fig. 56 illustrates a naked Crompton lamp. This lamp is of the differential type and has two solenoids, one in series and the other in shunt with the arc proper. Each solenoid is provided with a soft iron core, the upper ends of which are freely suspended from the opposite ends of a rocking lever. The cores will always hang vertically. Fig. 56 is a general view of this lamp and Fig. 55 is a diagrammatic sketch illustrating the principles of the working. In this latter sketch LL is the rocking lever and the cores at each end are shewn in dotted line. BW is a fixed *brake wheel*, and D is a drum fixed to it and turning about the same axis. BB is the brake band which is fixed to one end of the rocking lever and passes round the brake wheel to a spring S. A pair of flexible connectors are attached to the carbon holders, which are then suspended from the drum D, so that any movement of this will, according to the direction of the movement, draw apart the carbon holders or cause them to approach one another.

The spring S is so adjusted that with no current in the lamp the carbons will always run together. When a current is switched on the series core is pulled down. This turns the lever and pulls the brake band. The action of this tightens the brake and also turns the brake wheel and drum. This separates the carbons and the arc strikes.

As the arc lengthens the series current decreases and the current in the shunt coil increases. Thus the shunt core is gradually pulled down, and a stage is reached when the brake pressure just allows the carbons to feed together a little. Then the series current is increased and the brake holds the wheel. And so the action goes on. The brake is powerful enough to prevent any jarring together of the carbons when the lamp is subjected to violent vibration, yet it is sensitive enough to allow of an almost continuous and quite imperceptible feed of the carbons. A dash pot is seen on the left-hand side of the figure. This merely consists of a cylinder and piston in which there is a valve which will admit of the carbons feeding easily, but which prevents a too violent plunge of the series core when the lamp is switched on. This piston is also connected to the rocking lever.

Both poles of the lamp are insulated from the frame.

These lamps are supplied for any current from 5 to 20 amperes.

Energy absorbed in the Working of a Lamp. With one of the lamps described above, a 5-ampere lamp, the following data was obtained for the resistance, current, etc., of the various parts.

Part	Resistance	Current	Volts required	Watts
Series coil	0·5 ohm	5 amps.	2·5	12·5
+Carbon	0·1 ,,	5 ,,	0·5	2·5
Crater		5 ,,	39	195·0
Arc	0·5 ,,	5 ,,	2·5	12·5
−Carbon	0·2 ,,	5 ,,	1·0	5·0
Totals	1·3 ohms	5 ,,	45·5	227·5

It will be noticed that 41·5 of the total 45·5 volts and 207·5 of the total 227·5 watts are absorbed by the arc and crater. But the most striking item is the total resistance which is only 1·3 ohms. Of this the arc is stated to be 0·5 ohm, but this is the resistance of the arc when it is burning steadily with the normal distance between the carbons. As the carbons burn away the resistance increases, and this increase will considerably alter the value of the total.

Now it must be remembered that the crater always absorbs 39 volts, so that the current in the lamp is really maintained by difference between this and the applied E.M.F., and will be given by

$$\frac{\text{applied E.M.F.} - 39}{\text{resistance}}.$$

Thus in this case the current

$$= \frac{45\cdot5 - 39}{1\cdot3} = \frac{6\cdot5}{1\cdot3} = 5 \text{ amps.}$$

Use of Steadying Resistance. Now if the resistance of the arc proper should rise to 1 ohm the total resistance would become 1·8 ohms and the current strength would therefore drop to $\frac{6·5}{1·8} = 3·6$ amps.; and this is a considerable reduction, so much so that the lamp would most probably refuse to act properly with the result that the carbons would rush together. In this way steady feeding would be impossible, and the light would be variable, flickering, and annoying.

To counteract this a *steadying resistance* is put in series with the arc and a greater applied E.M.F. is used. With the lamp under discussion the steadying resistance supplied by the makers was 4·0 ohms resistance, and the E.M.F. to be applied was 65·5 volts. In this way the conditions in the arc were quite unaltered, the steadying resistance absorbing 20 volts with a current of 5 amps. Thus to the table above the following can be added.

Part	Resistance	Current	Volts required	Watts
Steadying Resistance	4 ohms	5 amps.	20	100
CORRECTED WORKING TOTALS	5·3 ohms	5 amps.	65·5	327·5

The introduction of this steadying resistance naturally lowers the efficiency, for now 207·5 of the total 327·5 watts are absorbed by the arc and crater. But at the same time the lamp will now work perfectly steadily and variations of the arc resistance do not produce such big variations in the total resistance. And further despite this deliberate though necessary reduction of efficiency the lamp will only absorb about 0·5 watt per candle power given out.

There is another important point in connexion with a steadying resistance. If there were none then when the carbons are in contact—preliminary to striking the arc—the resistance in the circuit would be extremely low, amounting only to 0.8 ohm in the case cited. Thus if the applied E.M.F. were 45.5 volts a current of almost 56 amperes would pass until the carbons were separated. This is not safe nor desirable, and it is seen that a steadying resistance is needed if only for striking the arc.

Candle Power Curves. The intensity of illumination from an arc lamp differs in different *directions*, and it differs also with the nature of the arc used. Figs. 57 *a*, *b*, and *c*, shew

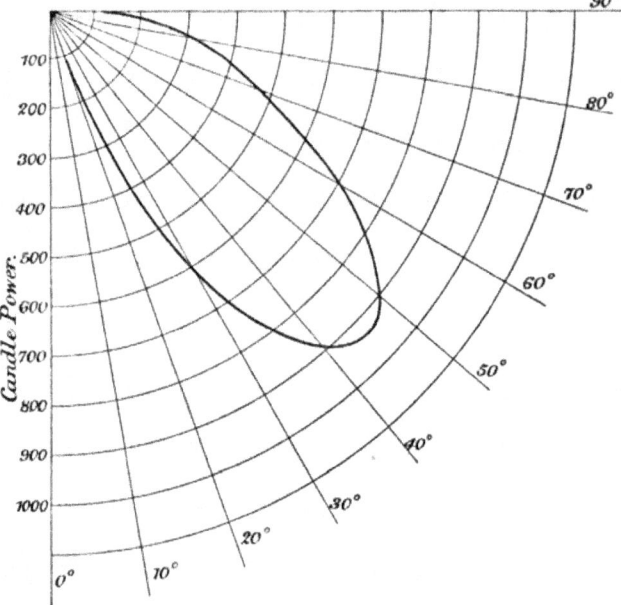

Fig. 57 *a*.

the relative distribution in different directions with open, enclosed and alternate current arcs respectively, of equal total C.P. lamps. With the open arc the maximum C.P. is obtained at an angle of 45° to the vertical. This is different in the case of the enclosed lamp; and with the alternate arc there is almost as great a c.p. in the upward directions as in the downward—the maximum in each case being less than with the other two, the light being more generally diffused.

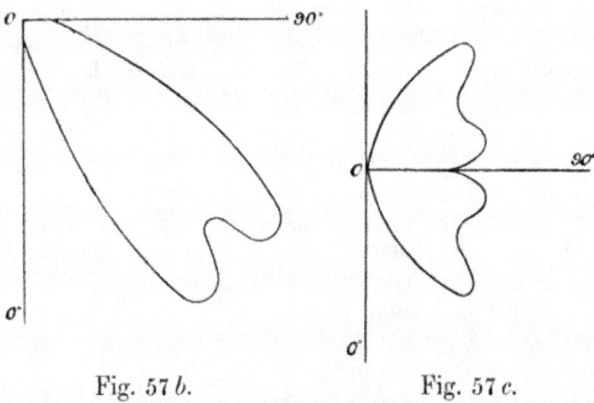

Fig. 57 *b*.   Fig. 57 *c*.

It follows that open and enclosed continuous current arcs should be fixed higher than alternate current arcs. In street lighting each arc usually illuminates an area of about 75 feet radius, the arc being placed at such a height that a line drawn from it to the edge of this area makes an angle of 77° with the vertical.

Enclosed Arcs. The lamp described above is known as an *open* arc, that is to say the arc burns in a globe which is not air-tight and consequently absorbs oxygen, whilst the carbons burn away.

There is a type of arc lamp in which the arc is produced in an air-tight globe, these lamps being called *enclosed* arcs. No attempt is made to exhaust the globe in the first instance, but the arc will soon absorb all the oxygen in it, and after that, no further supply being available, the carbons will no longer *burn away* although they will still be raised to the temperature of incandescence and the arc will still be capable of maintenance.

As a matter of fact it is impossible to get a strictly air-tight chamber and at the same time provide a mechanical arrangement for striking the arc, so that the carbons do burn away, but at a much smaller rate than with the open arc. Thus the labour of *trimming* the lamp (as the process of putting in fresh carbons is called) is reduced, and the amount of carbon used is also reduced to about one-seventh.

To counteract these advantages, however, the efficiency of enclosed lamps is not so high as that of open arcs, being about 1·5 watts per candle power against 0·5 to 1·0 watt per candle in the open arc.

The mechanical arrangements for striking and feeding do not necessarily differ very much from those employed in the open arcs except in the arrangement for feeding the top carbon into the air-tight globe.

Fig. 58 is a general view of the "W. J. Davy" enclosed arc lamp, and this does differ in many particulars from the usual two-coil "differential" gear employed. There is only one solenoid, the series, and the whole of the mechanism is so arranged that the variation of the resistance of the arc produces sufficient variation in the strength of the current in this coil to admit of smooth and regular feeding of the top carbon.

Owing to the absence of the uprush of cold air, such as takes place in the open arc and produces the curious and usual formation at the positive carbon, the carbons in the enclosed lamps burn with flat ends. As a result of this the light from the hotter positive carbon cannot get out. It is therefore necessary to *lengthen the arc*, and this can only be done by having a greater difference of potential. An E.M.F. of 60 to 75 volts is generally applied to the enclosed arc lamp terminals.

The candle power curve of an enclosed arc is shewn on Fig. 57 *b*.

In the lamp illustrated there is only one iron core I.

This is hung on one end of the beam and will be drawn down into the solenoid *F* when a current passes.   At the other end of the beam the rod *P* is the piston rod of a dash-pot *D*, which serves to steady the movements of the gear.   The rod

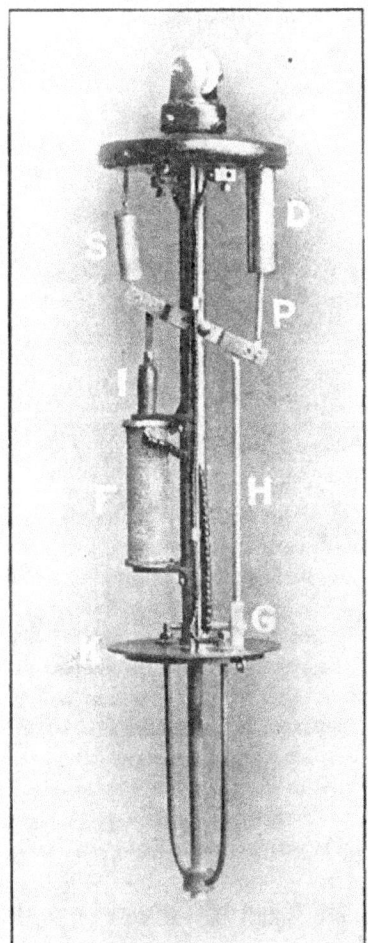

Fig. 58.

H is connected to the top carbon grip *G*, so that when the core *I* is sucked down into *F*, the top carbon will be raised up by this rod *H*. A spiral spring *S* is fastened to the top of the frame of the lamp and to the same end of the beam as the core *I*. Thus when the beam is depressed on this side the spring will tend to pull it up again.

The bottom carbon is not movable, all the feeding being done by the top. A globe, with only a top opening, is held by means of springs so that its mouth is pressed firmly against rubber washers on the under side of the platform *K*.

Thus it can be seen that when a current is passed the carbons will be separated. When the arc becomes a certain length the current will have decreased to such an extent that the spring *S* will be able to pull the core up somewhat, and eventually to pull it up enough to allow the bottom of the rod *H* to rest on the platform *K*. As soon as this happens the carbon holder allows the carbon rod to slip a little, and thus the resistance is reduced, the current is increased, the core *I* is drawn in again, *H* is pulled up again, and the carbon holder grips the carbon and raises it. The top carbon is a long one and it passes up the centre tube of the lamp, the grip holding it just at the platform *K*.

The grip is shewn in Fig. 58 *b*.

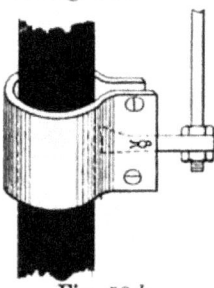

Fig. 58 *b*.

Arc Lamps in series. In practice it is generally found advisable from economic reasons to run two, three or four (according to the voltage of the power supply) arcs in series.

On the 100 volt supply, for example, a resistance would have to be used with every single lamp to absorb some 55 volts—and the product of this with the current strength would be absolute waste.

Hence it would be just as cheap to run *two* lamps in series. Each would require 45 volts, making a total of 90 volts, and a steadying resistance to absorb the remaining 10 volts would be all that was necessary. In this way for the same expenditure of energy twice the amount of illumination would be produced.

With a 220 volt supply, four lamps can be run in series ; and this proves more satisfactory than the case above since there is a wider margin, and a little higher steadying resistance ensures smoother working of the lamps.

Thus in a 220 volt service, arc lamps are arranged in groups of four in series, each group being, of course, in parallel with the others.

There is one difficulty to be overcome with arcs in series. If one lamp of a group breaks down, the other members of the group will cease to work. This is overcome by an automatic arrangement in shunt with each lamp. When a lamp breaks down a resistance is automatically put in which absorbs the same amount of energy as was being absorbed by the lamp. Thus the other members of the group are unaffected by the breakdown.

The principle of this automatically thrown-in resistance is illustrated by Fig. 59. A and B are the supply terminals to which the arc lamp is connected. A is connected to the terminal T_1 of the automatic resistance. From T_1 the circuit branches. One path lies through a fine wire coil Sh, and from this to T_2. The other lies through a resistance coil of platinoid wire which will absorb the E.M.F. necessary to work the lamp, and from that through a thick wire coil Sr. From this a connexion is made to a lever arrangement which can "rock" at H. A piece of iron I is fixed at right angles to this lever, and underneath the pole pieces of the iron cores of the two coils.

When the arc is burning properly the coil Sh is in shunt with it, and only a very small current indeed passes through. The coil Sr is not active under these conditions since there is no connexion between C and C'. But if the arc sticks and the carbons do not feed together the arc circuit will be broken,

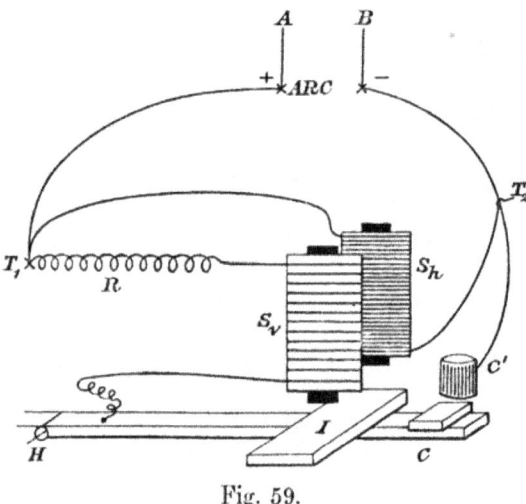

Fig. 59.

and when this occurs a larger current passes through the coil Sh. The effect of this is to draw up the iron I until the pieces of carbon at C and C' touch. This will immediately put the resistance R and the coil Sr in the circuit, and the other lamps of the circuit will not be affected.

Bastian Mercury-Vapour Lamp. This lamp is of the arc lamp type, the light being produced between two bodies of mercury instead of the two carbons. The mercury is enclosed in a sealed glass tube from which the air has been exhausted, so that the tube contains nothing but mercury and mercury vapour. At normal temperature there is a continuous thread of mercury from one end of the tube to

10—2

the other, each end consisting of a reservoir bulb. Electrical contact is made to the mercury by means of platinum electrodes fused into each of these bulbs. The form of the tube —it may be called the *burner* of the lamp—may be circular or S-shaped, and no special virtue seems to belong to any particular form up to the present.

When an E.M.F. is applied to the electrodes (through a resistance, for otherwise there must be a dead short circuit) and the tube is tilted so that the mercury thread breaks in the middle of the tube, an " arc " is formed between the ends of the break. This arc causes further vaporisation of the mercury, the pressure of which forces the threads back into the reservoirs. This increases the length of the arc, and the action goes on until the whole length of the tube is illuminated. All the time mercury is being vaporised and condensed on the walls of the tube.

Fig. 60 illustrates the mechanism and burner of the Bastian lamp. The mechanism is designed to tilt the

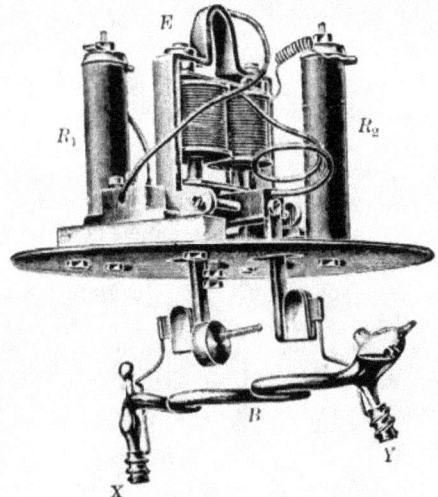

Fig. 60.

burner and so start the arc when the current is switched on. There is a form of mercury lamp in which this is done by hand, but obviously this would be disadvantageous for outdoor and much indoor lighting.

The burner is marked B, and is of double **S**-shape in this case. X and Y are the electrodes. The burner is hung on a frame lever which can be "rocked" by means of the electromagnet E. This is connected in *series* with the burner and the steadying resistances R_1 and R_2. When there is no current, the burner hangs in a horizontal plane. As soon as the current is switched on, the magnet becomes excited and draws up the lever arm against a small spring. This tilts the burner, the mercury thread breaks, and the "arc" strikes.

When the current is switched off the burner drops back to its normal horizontal position, and as it cools and the vapour condenses, the thread gradually fills up the tube and becomes continuous. Then the lamp may be started again.

The one drawback to the lamp is the character of the emitted light, which contains no red rays, and which is therefore disastrous for certain colour schemes. This may be remedied by running an ordinary carbon filament lamp in parallel with and adjacent to the vapour lamp. With this combination, the resultant light appears to resemble "daylight" very closely.

The efficiency is about 0·3 to 0·4 watt per C.P., as nearly as can be determined. The combined efficiency of a vapour lamp and the carbon lamp is about 0·7 watt per C.P. The life of the lamp varies from 3000 to 7000 hours, depending solely upon the glass of the burner. If it has been thoroughly annealed, an average life of 6000 hours may be expected.

The lamps are supplied for all ordinary lighting E. M. F.s, and the steadying resistance absorbs about $\frac{1}{3}$ of the voltage.

CHAPTER VII.

ELECTRIC INSTALLATION AND POWER DISTRIBUTION.

THE problems connected with the installation of electric light in a building are not very complex, but they are of great importance, and it is essential that everyone assisting in the work of installation shall know exactly what he is about and the reasons why he is doing a thing one way rather than another.

It will be advisable firstly to lay down the fundamental principles concerning wiring installations. A building is to be supplied with electrical power either for lighting purposes or for the purposes of driving machinery, or for both of these. This power may be bought from a supply company, or sometimes from the local Borough Council; or it may be generated by the tenants of the building themselves. But in all cases the power is supplied at a uniform pressure—a uniform E.M.F. This may be 100 volts or 200 volts, or indeed almost any desired E.M.F., but it must remain uniform. The current strength on the other hand will vary, and naturally will be proportional to the amount of electrical energy which is being absorbed by the consumer.

Impossibility of series running. Electric lamps are made of such resistance that when a definite E.M.F. is applied to them, a current passes whose strength is just enough to raise the filament to the temperature of incandescence. And lamps are made for 100 volt, 200 volt, 230 volt, etc., circuits.

Now a 100 volt lamp will only light properly with an E.M.F. of 100 volts. Thus if an electrical supply is 100 volts and it is desired to have a number of lamps in the various rooms of the building, it will be necessary to connect those lamps up in parallel with one another so that each lamp will have 100 volts E. M. F. applied to it.   If they be joined in series and the 100 volts applied to the ends of the series, it is obvious that no one lamp will have a sufficient E. M. F. to light it properly.

Fig. 61 illustrates the case with 3 lamps. D represents a dynamo running at an E. M. F. of 100 volts, and L, M, and N, represent three lamps in series connexion.   These are supposed to be 100 volt lamps, L being an 8 C. P., M a 16 C. P., and N

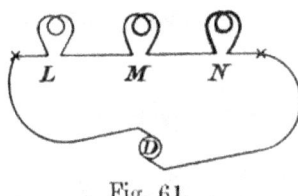

Fig. 61.

a 32 C. P.   Now all these absorbing equal watts per C. P., it follows that L will have the highest resistance, M the medium, and N the lowest; and that when applied to a 100 volt dynamo separately, the currents absorbed will be inversely as these resistances.

Now in the case illustrated by Fig. 61, the lamps are in series, and therefore the 100 volts is distributed down the three lamps.   Thus none of them will light up properly.

Again, the lamps being in series, it follows that the current strength is the *same in each*.   Since L has the greatest resistance and since the heating effect is proportional to $C^2 \times R$, it also follows that L (the small C. P. lamp) will be the hottest, and may light with a bright red glow. M will come next in point of temperature, and it may be glowing with a dull red, whilst N will in all probability not shew any sign of being heated.

This experiment serves the purpose of shewing that a lighting system of a number of lamps all joined up in series would be practically impossible. For even if it were possible to increase the E.M.F. of the dynamo to 300 volts in the case cited, the difficulty would not be overcome in the least since the fall of potential is proportional to the resistance. Hence the lamp L would get more than its needful 100 volts, and the lamp N would still get less. In a series circuit the current strength is the same at all parts.

The only thing to do would be to install lamps of equal resistance and C.P. Then there would be some possibility in the method. But as the number of lamps to be used is increased, the E.M.F. would have to increase also at an alarming rate : and would soon get beyond the bounds of comfort or safety.

Another point further—in a series system, if a lamp burned out the whole system would be cut off from the supply. It is thus seen that a series—or a constant current system—is a practical impossibility.

Parallel connexion of Lamps. But with a constant E.M.F. the lamps may be grouped in parallel, as in Fig. 62, where the same three lamps are shewn. $+ F$ and $- F$ are

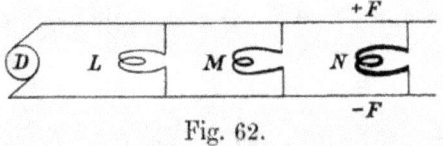

Fig. 62.

two wires or *feeders* from the terminals or brushes of the dynamo D. Between these feeders there is an E.M.F. of 100 volts, and it may be assumed *for the moment* that these feeders do not absorb any energy themselves and that therefore there is 100 volts difference of potential between them all the way along.

The lamps are connected as shewn across the feeders, and are thus in parallel with each other. Each lamp has a current in it just sufficient to light it properly, and this current

$$= \frac{100}{\text{Lamp Resistance when hot}}.$$

Thus each lamp has a different current in this case, the E. M. F. being constant; and the lamp N glows most brilliantly, since it is made to take a greater current, and has a greater surface area of filament at the temperature of white heat. But each lamp does just what it is designed to do.

Now more lamps can be joined across $+F$ and $-F$, and the E.M.F. of supply need not be altered. But of course since the lamps are all in parallel *the total resistance becomes less and less as more lamps are connected across the feeders*.

Just for the moment the reader might imagine that a big installation of lamps would therefore be a much easier thing to run than a small installation. But it must be remembered that the E.M.F. is to be constant—it *must* remain at this 100 volts, or whatever is chosen at the outset—and consequently as the total resistance of the installation becomes less and less with every added lamp the strength of the current must get greater and greater; and therefore the power absorbed (E.M.F. × current) must become greater.

It can be viewed from another standpoint. Each lamp requires a definite E.M.F. and a definite current. The E.M.F. is constant and every additional lamp means therefore an addition to the total current strength of the current required for that lamp. Thus as more lamps are used more total current is required.

Whether the dynamo can give this indefinitely or not is a matter which will be discussed in later chapters. It will be assumed at present that the supply is always equal to the demand. It is thus seen that a system of lamps in parallel is easily worked, for with a constant E.M.F. it is possible to use lamps of different candle power; the total resistance decreases as the number of lamps increases; and one lamp or more may burn out without affecting the other members of the installation.

Current strength at different parts of a parallel system. Before discussing the details of a typical installation, attention should be drawn to a very important point respecting the current strength in the different parts of the feeders. Consider Fig. 62 again. The total current strength will be

$$= \frac{\text{E.M.F. of dynamo}}{\text{Total resistance of hot lamps}}.$$

Now the dynamo resistance can be neglected in comparison with "outside" resistance—at any rate for the purposes of this discussion.

$$\therefore C = \frac{\text{E.M.F.}}{\dfrac{1}{\dfrac{1}{L} + \dfrac{1}{M} + \dfrac{1}{N}}}.$$

This current C will also be the sum of the currents absorbed by L, M and N, and will therefore $= \dfrac{E}{L} + \dfrac{E}{M} + \dfrac{E}{N}$, which is exactly the same as the expression above.

The feeders $+F$ and $-F$ have to transmit this current but it will be found that the total current strength is only carried between the dynamo and the first lamp. Let it be assumed that L requires 0·5 amp., M requires 1·0 amp. and N requires 2·0 amps. The total current will thus be 3·5 amps.

If ammeters be put in the feeder $+F$ at four places as illustrated in Fig. 63, then the readings will be found to be as

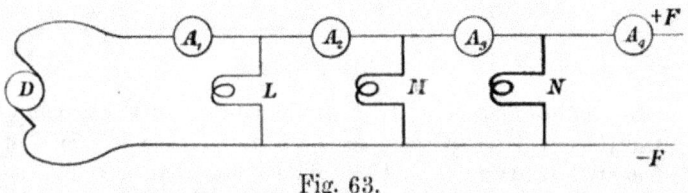

Fig. 63.

follows—$A_1 = 3\cdot5$ amps., $A_2 = 3\cdot0$ amps., $A_3 = 2\cdot0$ amps., $A_4 = 0$. This should be carefully thought out by the reader and compared with cases of series circuits.

Thus the feeders do not carry the total current throughout their whole length : and in practice this fact is utilised to cut down the area of cross-section of the feeders of the various branches of an installation in order to reduce the initial cost of material.

Determining points on the sizes of cable to use. The feeders are bound to absorb some of the energy of the system—the amount being determined by the square of the current strength and the resistance. Clearly the current cannot be altered if a definite amount of energy at a definite pressure is required. Therefore the resistance of the feeders must be kept as low as possible. But this will mean larger feeders, and greater cost of installation. Further, no feeders can be obtained which shall offer no resistance. A point will occur then when it will be cheaper to waste a certain amount of energy in these feeders than to pay interest on the extra capital expended in putting down larger cables.

But there are other determining points. Since energy is absorbed by a cable then there will be a difference of potential between one end and the other—there will be a "drop in volts" along the feeder which will be determined by the product of the current strength in, and the resistance of that cable. For example, in Fig. 64 the dynamo is some distance

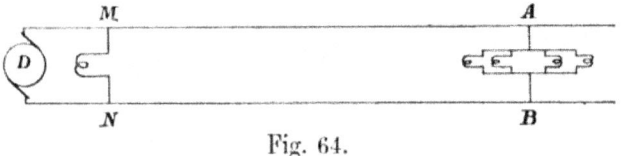

Fig. 64.

from the installation at AB. If the lamps be 100 volt lamps and the dynamo is running at 100 volts then the difference of potential between A and B will be less than 100 by the amount of the "drop" along the cables. There might also be lamps at MN near to the dynamo. Now if the dynamo runs to keep these lamps right the others will be underrun. If on the other hand the lamps at AB are kept at 100 volts those at CD will be overrun. It has been already pointed out that it is bad policy to underrun or overrun lamps.

Hence the feeders shall be of such resistance that with the maximum load being transmitted there shall not be a greater drop between the nearest and the furthest lamp than 2 °/₀ of the mean pressure at which the lamps are run. This has been laid down by the Institution of Electrical Engineers as a definite regulation of wiring installations.

The last point to consider with regard to the size of the feeders is the heating effect. There must not be undue heating—both from the point of view of safety and of economy. The official regulation on this matter allows a *maximum temperature of 30° F. above the temperature of the surrounding space.*

A Fundamental scheme of installation. Before working out examples on all these questions, the planning of an installation may be advantageously considered.

It can be seen that an installation could be laid in a building by simply running the pair of feeders through every room and corridor and connecting such lamps to those feeders as may be considered necessary. Each lamp could be controlled by a separate switch and fuse, or groups could be controlled by a common switch.

Fig. 65 is a diagram of such an installation. *D* is the

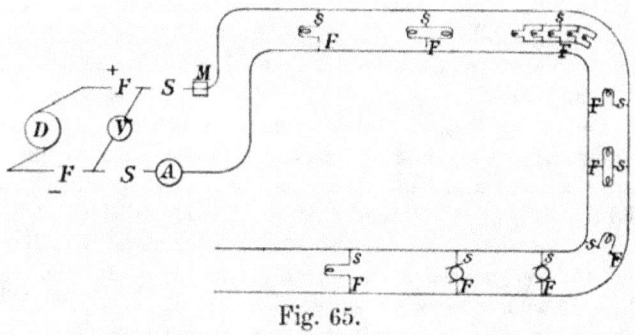

Fig. 65.

dynamo or the source of power. $+ F$ and $- F$ are two *main* fuses which will "blow" should the current ever become greater than the maximum for which the circuit is designed. S and S are two switches—or one "double-pole" switch. A is an ammeter for measuring the total current, V a voltmeter to measure the E.M.F. of the dynamo, and M a meter for measuring the total energy consumed. This may be an ampere-hour meter or a watt-hour meter (see ch. XVI).

The feeders are carried round the building and lamps or motors are connected to these where desired.

Switches are shewn by s; fuses by F; lamps by the customary loop; and motors by $\mathbb{C}$.

In this diagram are seen single lamps, groups of two and more on one switch, and motors. Now this is not the scheme adopted in practice except for very small installations.

Generally adopted scheme. It would not be economical to run the big feeders all round a building—neither would it be a simple matter. Further a breakdown on the feeder would be expensive to repair, and would perhaps inconvenience the whole of the system.

The result is that the system is divided up into branches, and *distributing feeders* or *distributors* are run from a common distributing centre—known as the *switchboard*—to the various branches.

For example it might be supposed that a building is three storeys high. Each floor can be taken as a branch. Or the building might be straggling with front and back wings. In this case each floor would be divided into two branches—front and back.

Fig. 66 is a diagram which serves to illustrate the approved method of distribution of electrical power. The dynamo, main fuses, switches and meters are shewn as before and the *main feeders* are then connected to two copper bars $+B$ and $-B$ on a switchboard. These bars are known as *Bus bars*, and are merely used as a convenient means of distribution. On the diagram shewn three sections or branches are taken from the bus bars—but of course any number may be taken on the same principle. S_1 and S_1' are the switches controlling

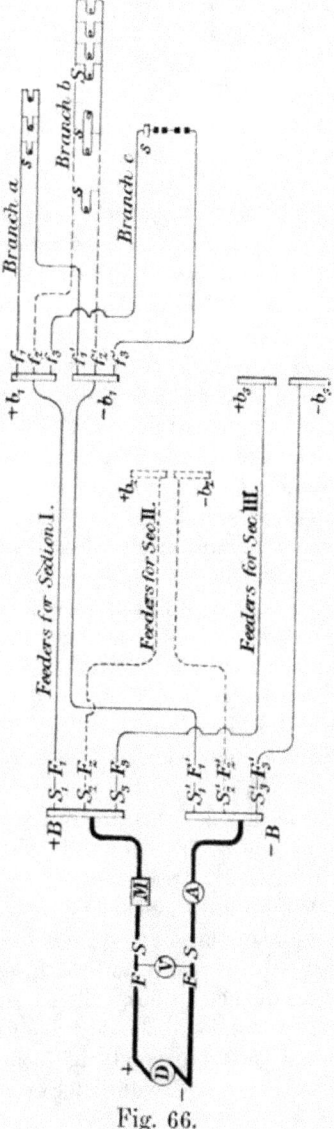

Fig. 66.

the whole of section I, and F_1 and F_1' are corresponding fuses for that section. The feeders for section I are shewn by firm lines. These feeders are smaller in area of cross-section than the main feeders between the dynamo and switchboard—and their size will depend upon the strength of the current required for the section. These feeders are connected to two smaller copper bus bars $+b_1$ and $-b_1$. These bars are on another board known as a *distributing board*—and this is placed at some central position for the distribution of current to its own section. The distributing board does not usually have switches on it, but it has double-pole fuses—that is to say a fuse is put on each wire of every branch going from the board. f_1 and f_1' are the fuses for branch a, which is shewn to consist of three lamps controlled by one switch. These will be in one room—or along a corridor—the switch being placed near the door or in some readily accessible place. Similarly there is shewn branch b, consisting of a single lamp and switch, two lamps on one switch and four lamps on one switch. Branch c shews two arc lamps in series with one switch and a steadying resistance.

It can be seen that should anything unforeseen occur on any one of the branches, such as a short circuit, only the fuse governing that branch would be blown—and the rest of the installation would not suffer. Or should a short circuit occur on the feeders for section I, only fuses F_1 and F_1' would go.

The rest of this system may be filled in easily by the reader. S_2, S_2', and F_2, F_2' are the switches and fuses for section II, and the feeders for this are run to another distribution board, from which branches are taken as desired. And section III is similarly treated.

On the main switchboard there should be placed the main fuses, and switches, the ammeter, voltmeter and *watt-hour meter* if one is used; the bus bars; and the fuses and switches controlling the whole of the sections. This board is generally placed in the dynamo-room and should be so fixed that there is easy access to the back and plenty of room behind it for first wiring and for alterations, additions or repairs.

Each distribution board should be placed at some central position in the section for which it distributes, preferably in a corridor, about 5 feet from the ground, so as to be readily got at in the event of a fuse being blown, and should be provided with a locked glass cover.

The system described above illustrates what is known as the "two-wire system" of distribution, a cable being used for both "out" and "return."

One-wire system. Fig. 67 illustrates what is called the "one-wire" system of electric lighting, a system which is used largely in the wiring of ships. The negative brush of the dynamo is "earthed"—in the case of a ship, is connected to the bulkhead. There is only one main feeder, to one bus bar on the switchboard. From this bar feeders are taken for the various sections—one feeder to each section. The distribution board of each section consists of one bus bar and fuses for the different branches as shewn in the figure. The connexions to earth or to the bulkhead of a ship are shewn at all places by the arrow-head and the letter E.

This system is cheaper to install than a two-wire system. There is less wire to be used, less work in the wiring, and smaller switchboard and distributing boards. But the circuits are only protected by single pole fuses and if any part of the positive side, that is any part of wiring should become "earthed"—connected to earth or to the bulkhead—then a dead short-circuit is the inevitable result. The result of that one cannot predict, but it can never be pleasant. At the same time one is bound to say that the system is run without trouble—according to owners at any rate—on several large lines of steamers. But on the other hand there are just as many if not more with a two-wire installation.

On land the one-wire system would not be allowed by electrical supply companies—and probably not by insurance companies.

Three-wire system. The "three-wire system" is utilised on direct current installations when it is desired to have two different E.M.F.S ; and it involves the use of two dynamos.   It is rarely used in private installations ; but many electric

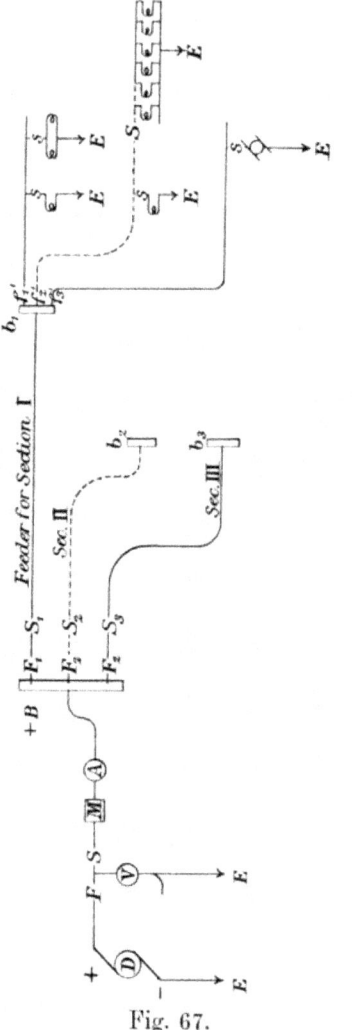

Fig. 67.

power supply companies use it for their distribution as it
saves them some cost in feeders because they can use three
mains instead of four.

Fig. 68 illustrates the main lines of the system. D_1 and
D_2 are two dynamos joined up in series as shewn. From the
+ brush of D_1 one main feeder is taken. The – brush of D_1 is

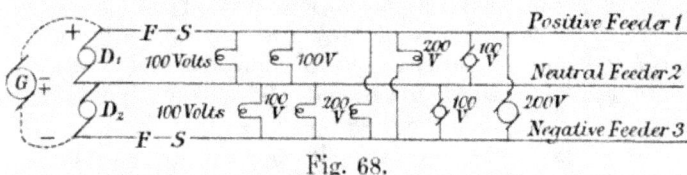

Fig. 68.

connected to the + brush of D_2 and from this common connexion a second main feeder is taken ; and from the – brush
of D_2 the third feeder as shewn.

Let it be supposed that each dynamo has an E.M.F. of
100 volts, then the total E.M.F. between feeders 1 and 3 will
be 200 volts. The " middle " wire (feeder 2) is sometimes
earthed so that its potential is zero. This wire is generally
known as the neutral wire.

Now between feeders 1 and 2 or feeders 2 and 3 there is
available an E.M.F. of 100 volts. 1 is positive to 2 ; and 2
is positive to 3. Hence between feeders 1 and 2 an installation of 100 volt lamps etc. can be supplied ; and the same
applies to the feeders 2 and 3. Between feeders 1 and 3 a
200 volt system can be installed. Here then is the first
advantage of the system—two E.M.F.s available. The second
and last advantage of the system lies in the fact that only
3 feeders are being used for supplying two separate 100 volt
installations. For if D_1 and D_2 were independent machines
supplying 100 volt installations separately 4 feeders would be
necessary on the general two-wire scheme.

Furthermore, the middle or neutral wire in the three-wire
system only carries the difference of the total currents in two
100 volt installations, and in an ideal system the two installations are so arranged that they take equal currents—it is said

that they "balance"—with the result that the middle wire does not convey any current at all. This of course applies to the perfect and ideal installation ; but in practice it is quite impossible to keep a perfect balance continuously since a consumer may switch lamps on or off as he pleases. Hence it is necessary to have a middle wire—but mainly as a half-way *potential* point between the two outer wires of the system; and to carry such difference of current as may exist between the two 100 volt installations. But this difference will always be less than the current strength in either of the "outers," consequently the middle wire need not be so large in area of cross-section as either of the outers, it being usually one-half.

The disadvantages of the system tend to outweigh its advantages—although that may depend upon the point of view of the user. Firstly unless the two sides of the system are balanced one dynamo will have more work to do than the other, with the result that one will tend to send current through the other which will in turn tend to drive it as a motor. The actual result will be that the speed of one dynamo will be altered, and consequently a different E.M.F. will be yielded to the general inconvenience of the installation on its side of the system.

To overcome this disadvantage various plans may be adopted. One consists in putting in an artificial load on that side of the system which is taking the lesser current until a balance is produced. This of course may also mean waste, unless the artificial load is a useful one—such as charging secondary cells. It also means the almost constant attention of someone to see that the balance is preserved.

Another plan is to have a portion of the installation so wired up that it can be changed over from one side to the other—that is to say a set of lamps arranged that they can be switched over from feeders 1 and 2 to feeders 2 and 3. This of course will involve some initial outlay in wiring which will tend to reduce the advantage derived by the saving of the fourth feeder.

Generally however some plan is adopted after the lines

11—2

illustrated by Fig. 68 with dotted connexions added. ˙ G is a large generator capable of supplying the full load on the outside feeders. D_1 and D_2 are then coupled together—they are relatively small machines—and connected as shewn.   The system D_1 and D_2 will then be a sort of motor generator, or *motor booster*, or *equaliser*, or *balancer*. For, by means of these when there is excess of current on one side that machine will run quicker and the other will therefore follow it and raise the E.M.F. in order to preserve a balance.   If the excess is on the other side the machines will reverse their functions.

It is common for these to be shunt wound, with the field of each connected to other side of the system of that to which its armature is connected.   It must be understood that the main load is yielded by G, D_1 and D_2 being used merely as balancers.   There are other plans of balancing, but the general principle is the same.

A small consumer—e.g. a householder—buying electrical power from a company who have a three-wire system, does not experience any disadvantages, because all his lamps are connected up on one side of the system.   His load is balanced perhaps by that of his next door neighbour who may be supplied from the other side of the system.

But a large consumer is frequently put to much initial expense, because the company may insist upon his installation being balanced on itself.   This means that his installation shall be divided up into two equal parts—one of which shall be supplied from feeders 1 and 2, and the other from feeders 2 and 3.   This means a more elaborate switchboard and the initial cost of such an installation will be more than that of a similar two-wire system.   This of course is merely a disadvantage to the consumer.

Installing for alternate current supply.   So far it has been assumed that only continuous or direct current has been distributed, but the principles laid down for wiring installations hold equally for alternating currents—currents which are being reversed in direction some 50 to 120 times per second.

However, on alternating current systems the one-wire and three-wire schemes are never used, the two-wire standing alone. The diagram (Fig. 66) holds equally for alternating or continuous currents. If the installation is a private one the dynamo *D* will be an alternator—that is, a dynamo which yields an alternating current. If the installation is supplied with power from a company a *transformer* will take the place of the dynamo in his wiring scheme. The functions of a transformer will be discussed later, but it may be mentioned here that power supply companies supplying alternating currents generally run their alternators at a high voltage—say 1000 or 2000 volts—and this is *transformed down* to *any* desired E.M.F. at the ends of a consumer's installation.

High and Low Tension Distribution. This raises a very interesting and important matter in connexion with power distribution. It must always be borne in mind that the electrical power supplied is measured by the product of the E.M.F. and the current, and that the total electrical energy is the product of this and the time for which it is continuously supplied.

Hence it is seen that a given power may be supplied either by a small current at a large E.M.F., or a large current at a small E.M.F.

Now supply companies prefer to distribute a given amount of energy per sec.—that is, a given power—by transmitting as small a current as possible at as large an E.M.F. as is practicable for the very good reason that with a small current smaller main feeders can be used, and the waste of energy in the feeders can be decreased. This means more efficient distribution at a possibly smaller initial outlay in installation. On the other hand, of course, the feeders will have to be extremely well insulated since there is so large an E.M.F. between them.

But no ordinary electric light installation can do with a

greater E.M.F. than 230 volts, and even that is not so good as 100 volts, chiefly owing to the fact that better lamps can be made at present for the lower voltage than for the higher. Hence it is necessary that the high E.M.F. of production shall be transformed down to 230 volts or less at the ends of a consumer's installation. An alternating supply can be transformed down to any desired E.M.F. by means of stationary transformers—which do not require any attention, and of which the very existence is often unknown by the consumer.

But on a continuous current supply there is no such simple means of transformation; hence it is that companies supplying continuous current have to run at the same E.M.F. as that which their consumers can use without transformation. Their's is a low-tension system of distribution. This is not so economical to run as a high-tension system, and that is the reason why so many power companies have an alternating current supply.

Change-over schemes. An installation may be supplied by a consumer having his own engine and dynamo. But this only pays when the total load exceeds 20 kilowatts. When the load is less than this it will be more economical to "switch-over" on to a company supply. Again it will be useful to be able to "switch-over" at will, because one's private engine might break down. The company would then act as a "stand-by"—and would at the same time take the light loads and obviate the necessity of running the engine under wasteful conditions, since neither an engine or a dynamo run at their best efficiency unless running at their full load.

There are sometimes difficulties about changing over. For example, a private supply might be continuous current, whilst the local company supply was alternating. A change over from one to the other would be somewhat inconvenient unless the measuring instruments used were adapted for both currents; and the installation a simple one of incandescent lamps. Arc lamps and motors designed for continuous current will not work with alternating current—and *vice versa*.

Again, if both are similar supplies, the E.M.F.s must be the same : and this means that the private consumer must get his dynamo and installation to work at the same E.M.F. as that of the supply which he is going to use as a stand-by.

Further, he may be going to switch over to a 3-wire system on which he has to balance his full load. This means that his installation shall be divided up into two parts and that his switch board shall be arranged that on his own dynamo he can work a 2-wire scheme and on the supply a 3-wire scheme. All these matters must be attended to before the private installation is commenced ; or the cost and annoyance of necessary alterations will be excessive and wasteful.

Fig. 69 illustrates the general idea of a change-over arrangement. A and B are the "ends" of an installation. These are the middle points of a change-over switch—an arrangement by means of which A can be connected to H or

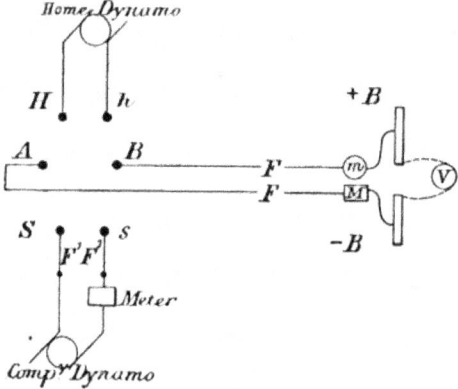

Fig. 69.

to S, and at the same time B is connected to h or to s. To H and h the terminals of the home dynamo are connected ; and to S and s the terminals of the supply dynamo—the stand-by—are connected through two fuses and through a meter. These, fuses and meter, are the property of the supply

company, and the consumer is not allowed to tamper with
them. The fuses are sealed, and should they be blown the
company will have to be communicated with before connexion
can be restored. The meter will measure the amount of energy
which is being taken from the company's mains.

Fig. 70 illustrates a more complicated change-over
scheme, to wit from a 2-wire home supply to a 3-wire
company supply.

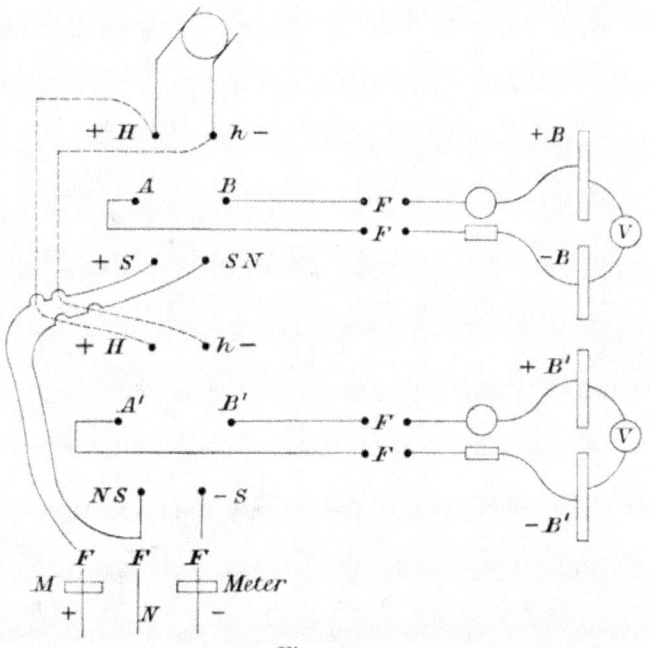

Fig. 70.

Two change-over switches are needed: and the installation must be divided into two sections. This means four
main bus bars on the switchboard, and these are shewn by
+ B, – B, + B' and – B' on the figure.

AB and A'B' are the ends of the two sections of the
installation. + H and – h are joined to the home dynamo

terminals as shewn, so that when the home supply is in use
the two sections are joined up in parallel ; which means in
other words that $+ B'$ is connected to $+ B$ and $- B'$ is connected to $- B$.

The three feeders from the company's supply are brought
in as shewn through three main fuses and two meters, one on
each "outer." $+ S$ and Ns are connected to the $+$ and Neutral
feeders ; and NS and $- S$ to the Neutral and $-$ feeders respectively.

Thus when the ends AB and $A'B'$ are connected to
$+ S$ and Ns, and to NS and $-s$ respectively, the result is
that $- B$ and $+ B'$ are connected together—and between
$+ B$ and $- B'$ there will be an E.M.F. of double the working
E.M.F. of the home supply.

In these change-over systems care must be taken that in
the change the bus bars still retain their relative potentials.
That is to say the bar $+ B$ must always be positive to bar
$- B$; and $+ B'$ always positive to $- B'$.   The reason for this
is that firstly the instruments may only read for one direction
of the current through them ; secondly, if there be any arcs
in the system the carbons would become reversed in their
burning if the current was reversed ; and, thirdly, it is always
desirable to work methodically even if for no other cause than
for symmetry of arrangement.

Systems of fixing Leads and Cables.   There are
several methods of fixing cables and fittings in an electric
light installation, but the details of these are generally entered
into in special courses on wiring and fitting, and this volume
does not make any attempt at specialisation.   It will be well
merely to give passing mention to the chief methods of cable
fixing.

A company's main feeders are generally run underground
in special "ducts" of wood or of earthenware and are covered
over with bitumen which is both damp proof and insulating.
They may be run in earthenware pipes, or in steel pipes ; or
they may be lead-cased cables, in which case they need not be

run in any prepared duct ; or they may be armoured cables, protected, that is to say, by an iron sheathing which in turn is covered with a damp proof preparation.

In a building feeders and cables may be fixed with "cleats" —pieces of wood with two grooves about one and a half inches apart—which serve to keep the wires in position along a wall or ceiling, and at the same time keep the wires apart from and parallel to one another. These are spaced at intervals, depending upon the size of the cables, from one another. This method is only for use in dry places and the wires must all be visible. It is the cheapest method of fixing, but is not often allowed either by the supply companies or by the insurance companies interested.

A second method consists in using wooden casing, by means of which all the wires are boxed in. Casing consists of long strips of wood with two grooves ploughed along the length parallel to one another and to the edge of the wood. The cables are laid in these grooves—one in each—and a thin wood covering, known as the *capping* is *screwed* on the top, so boxing in the entire length of cable. This is a very good system provided that it is done well.

A third system consists in running all the cables through thin steel tubes which may be fixed on the walls and ceilings by means of saddles or which may be built into the walls, etc., whilst a building is being erected. This system has come very largely into vogue in recent years. Its disadvantages lie in the facts that the two wires (or more) are not kept separate in the tube, that the insulation of the cable may be damaged in the "drawing-in" process, and that the cables cannot be exposed to view as they can with the casing and capping system. Further, it is necessary that precautions should be taken to prevent moisture getting into the tubes and there condensing ; and the tubes should be well earthed.

There is no doubt that for some purposes the "draw-in" system, as this is often called, is preferable to the casing, but on the other hand the casing possesses important advantages over the tube.

Insulated tubes can also be used, and whilst these admittedly yield excellent results they are nevertheless expensive to install. Special fittings are made for these tube conduits—such as bands, **T** pieces, elbows, unions, and inspectional boxes.

If the reader will. refer back to Fig. 66 and look at the arrangement of *Branch a* of *Section I*, he will note that three lamps are shewn controlled by one switch. The connexions from a lamp to the feeders of the branch may be made by *jointing* smaller wires to the main wires and running the smaller wires to the lamp. And there are other cases where connexions could be made by means of *joints*. The feeders themselves might be made up of two or more lengths jointed up to make them long enough. But wherever it is possible joints should be avoided—and certainly there should be no joints in a main feeder, or between any distribution board and the main switchboard.

A joint may offer a greater resistance—length for length—than any other part of a cable. Hence it would become hotter, and the insulation of a joint might not be as good as that of the rest of the cable. It is for these reasons that joints should be avoided wherever possible.

At the same time, if a joint be thoroughly well made and properly insulated there can be no objection to it ; but the fact that nowadays joints are almost entirely tabooed reflects upon the workmanship of wiremen in general.

To dispense with joints, a scheme known as *looping-in* is employed. This entails, perhaps, less labour than jointing, but it involves the use of much more wire in installation. On this scheme *Branch a* referred to above would be wired as shewn in Fig 71, and *Branch b* as shewn in Fig. 72. In this

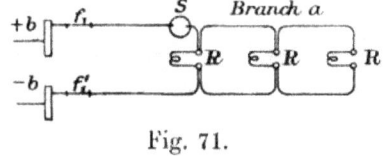

Fig. 71.

looping-in scheme all new contacts are made in the fittings. *Branch b* is typical of a more general scheme than is *Branch a*.

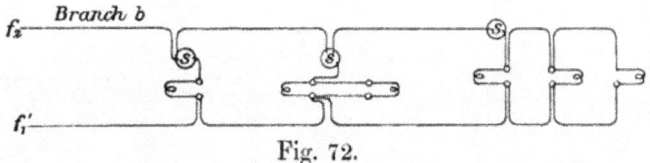

Fig. 72.

Wires and Cables. All wires and cables used in electric light installations are made of copper. This is the most economical substance to use, for wires made of other substances of equal conductivity and equal length would at the present market rates prove more expensive.

These conductors have generally a circular cross-section —but this is merely a matter of convenience and does not give any better conductivity than an equal cross-section of any other shape. The sizes of these wires may be given as the inches or centimetres of their diameter ; or more generally according to a standard wire gauge, in which case the number of the gauge denotes the size.

A conductor may consist of a single wire, or of a number of wires. A conductor of single strand is called a *wire* ; and one of many strands is called a *cable*. Thus a cable may consist of seven strands of a number 18-gauge wire ; and it is spoken of as being a $\frac{7}{18}$ cable, the numerator giving the number of strands and the denominator the size of each strand.

In electric installations *wires* are never used. Instead of using a single 18-gauge wire, a $\frac{3}{22}$ cable would be used. This has approximately the same area of cross-section and the same conductivity, but is used in preference to the wire because firstly it is *more flexible* and secondly it has a smaller chance of a complete break, due to being brought round a bend or similar cause.

The size of cable to be used must depend entirely upon the strength of the current it is to carry and upon the length to

Size	Diameter		Area		Resistance		Minimum Insulation Resistance. Megohms per mile	B. of T. amps.	Max. current allowable. I.E.E.
s.w.g.	Inches	Cms.	Sq. inches	Sq. cms.	Ohms per 1000 yds.	Ohms per kilometre			
22	·028	·0711	·0006158	·00397	39·05	42·7	2000	0·6158	1·7
20	·036	·0914	·001018	·00657	23·62	25·83	2000	1·018	2·6
18	·048	·1219	·001810	·01168	13·28	14·53	1200	1·810	4·2
16	·064	·1626	·003217	·02075	7·478	8·18	1200	3·217	6·8
14	·080	·2032	·005027	·03243	4·784	5·23	800	5·027	9·8
3/22	·06	·1524	·001825	·01177	13·18	14·41	1200	1·825	4·26
3/20	·078	·1981	·003016	·01946	7·97	8·718	1200	3·016	6·44
3/18	·102	·2616	·005364	·03461	4·48	4·902	1200	5·364	10·31
7/22	·084	·2314	·004266	·02752	5·636	6·164	800	4·266	8·54
7/20	·108	·2743	·007152	·04550	3·410	3·729	600	7·15	12·9
7/18	·144	·366	·01254	·0809	1·918	2·097	600	12·54	20·68
7/16	·192	·488	·02227	·1437	1·080	1·181	600	22·27	33·12
19/20	·180	·4572	·01912	·1234	1·257	1·375	600	19·12	29·23
19/16	·320	·8128	·06039	·3896	0·3981	0·435	600	60·39	75·06
19/14	·400	1·016	·09442	·6091	0·2547	0·278	600	94·42	108·3
37/16	·448	1·138	·1176	·7587	0·2045	0·224	400	117·6	129·6
61/16	·576	1·463	·1939	1·251	0·1240	0·1356	300	193·9	195·4
91/14	0·880	2·235	·4519	2·915	·0532	·0936	300	451·9	391·0
91/12	1·114	2·906	·6738	4·347	·0315	·0344	300	673·8	542·5

be used. Overheating has to be guarded against : and the "drop" in volts has not to exceed 2 °/₀.

A table is given on page 173 which shews the size of a cable in gauge, the diameter of a single strand in inches and centimetres, its cross-sectional area in square inches and in square centimetres.

Diameters of cables and thicknesses of insulation are often given in *mils*. A *mil* is $\frac{1}{1000}$ part of an inch and must not be confused with a *millimetre* which is $\frac{1}{1000}$ part of a metre.

The area of cross-section of a circular section wire is sometimes given in *circular-mils*. A wire of diameter 1 *mil* ($\frac{1}{1000}$ inch) is said to have a cross-sectional area of 1 *circular-mil*. This is taken as the unit. Hence a wire of 8 *mils diameter* will have an area of 64 *circular-mils* : that is to say its cross-sectional area is 64 times greater than the cross-sectional area of a wire of 1 mil diameter and 1 circular-mil area of cross-section.

The Board of Trade regulation allows a current of 1000 amperes per square inch cross-sectional area of copper cable, and the cross-sectional area of a cable may always be worked out on this basis.

The Institution of Electrical Engineers makes a variable allowance, their regulation being that "the size of conductors within a building will be determined by the permissible drop in volts which should not exceed 2 per cent. on lighting circuits."

This regulation allows a greater current for a given sectional area on small cables than the uniform 1000 amperes per square inch rule of the Board of Trade, and a slightly less current for the large cables. This is seen by consulting the table on page 173.

Worked Examples. Some examples may now be worked out illustrating all these points.

1. What is the resistance of 50 yards of $\frac{1}{16}$ copper cable, being given that the diameter of a single 16 wire is ·064 inch,

and the specific resistance of copper is ·00000066 ohm per inch per sq. inch?

To get first the resistance of 50 yards of 1 strand of 16 copper. It has been shewn (page 60) that

$$r = \frac{s \times l}{a},$$

where $r =$ resistance in ohms of a length l having a specific resistance s and an area of cross-section a.

And again $a = \pi \times r^2$, where $r =$ radius of the strand.

$$\therefore r = \frac{\cdot 00000066 \times (50 \times 36)}{\pi \times \cdot 032 \times \cdot 032}$$

$$= 0 \cdot 3695 \text{ ohm.}$$

It will be noted that l has been put into inches, since a will be square inches, and s is in ohms per inch per square inch.

Now having got the resistance of one strand of 50 yards, the resistance of the whole cable may be regarded as that due to seven equal resistances in parallel. If R denote the combined resistance, then

$$R = \frac{1}{\dfrac{1}{r_1} + \dfrac{1}{r_2} + \dots \dfrac{1}{r_7}},$$

but as these are equal to one another and to r determined above,

$$\therefore R = \frac{r}{7} = \frac{0 \cdot 3695}{7} = 0 \cdot 0528 \omega.$$

Now it has been found that the *resistance of a stranded cable is always higher, by 3 %, than the resistance of a solid wire of the same cross-sectional area and length of copper.*

Hence the correct practical value for R will be

$$\cdot 0528 + \left(\frac{3}{100} \times \cdot 0528 \right) = 0 \cdot 0528 + \cdot 0016 = 0 \cdot 0544 \text{ ohm.}$$

In a 1000 yards of stranded cable each individual strand will be longer than 1000 yards, hence the increase in resistance.

2.   To determine the resistance of two cables coupled in parallel, one being a 90 yard length of $\frac{7}{18}$ cable and the other a 150 yard length of $\frac{19}{14}$ cable.   It is given that 1000 yards of a single 18-wire has a resistance of **13·28 ohms**; and 1000 yards of a single 14-wire has a resistance of **4·784 ohms.**

90 yards of a single 18 has therefore a resistance $= \frac{90}{1000}$ $\times 13\cdot28 = 1\cdot1952\omega.$

$\therefore$ 90 yards of $\frac{7}{18}$ has a resistance $= \dfrac{1\cdot1952}{7} = 0\cdot1707\omega.$

150 yards of single 14 has a resistance $= \frac{150}{1000} \times 4\cdot784$ $= 0\cdot7176\omega.$

$\therefore$ 150 yards of $\frac{19}{14}$ has resistance $= \dfrac{0\cdot7176}{19} = 0\cdot0378\omega.$

The combined resistance R of these two in parallel

$$= \frac{1}{\dfrac{1}{0\cdot1707} + \dfrac{1}{0\cdot0378}} = 0\cdot0314 \text{ ohm.}$$

3.   Two main feeders of copper are to be used to convey electrical energy from a dynamo to a house 50 yards away.

The maximum current required will be 160 amperes and the E.M.F. at the house is to be 100 volts.

Allowing a 2 % drop in volts along the feeders, what resistance must they have, and hence what must be their size?

Since a 2 % drop is allowed therefore the dynamo can run at 102 volts—hence the total drop along the feeders will be 2 volts.

That is to say 1 volt drop along the 50 yards "out" and 1 volt on the 50 yards "return."

Hence the problem may be treated as though a single cable of 100 yards was under consideration.

The maximum current in the cable = 160 amps.

The difference of potential between its ends when this current is passing = 2 volts.

$\therefore$ The resistance of the cable $= \frac{2}{160} = $ **·0125 ohm.**

Now the area of cross-section can be calculated, knowing that

$$\text{Resistance of a length} = \frac{\text{specific resistance} \times \text{length}}{\text{area of cross-section}}.$$

$$\therefore a = \frac{s \times l}{R}.$$

$$\therefore a = \frac{\cdot 00000066 \times (100 \times 36)}{\cdot 0125}$$

$$= 0 \cdot 1901 \text{ square inch.}$$

And by consulting a table the cable which has the nearest cross-sectional area to this will be the one to be used, and its wire gauge size can be read off.

Thus on consultation it is seen that a $\frac{61}{16}$ cable has an area of cross-section of $0 \cdot 1939$ square inch. This is the nearest size *greater*, and is the cable which must be used.

The result could have been got from the cable when the resistance of 100 yards was determined.

For example, 100 yards has a resistance $= \cdot 0125$ ohm.

$$\therefore 1000 \text{ yards} \quad ,, \quad ,, \quad = 0 \cdot 125 \text{ ohm,}$$

and from the table it is found that a $\frac{61}{16}$ cable has a resistance of $0 \cdot 1240$ ohm per 1000 yards.

4. A group of 25 lamps each 16 c.p., each absorbing 3·5 watts per c.p. and running at 100 volts are to be fed by a dynamo 40 yards away. What must be the cross sectional area of the cable used if the drop is not to exceed 0·5 volt ?

Watts absorbed by each lamp $= 3 \cdot 5 \times 16 = 56$ watts.

Total watts absorbed $= 56 \times 25 = 1400$ watts.

$\therefore$ The current strength $= \frac{1400}{100} = 14$ amps., since watts $=$ E.M.F. $\times$ current.

$\therefore$ The resistance of the cable which shall require 0·5 volt to maintain a current of 14 amperes along it

$$= \frac{0 \cdot 5}{14} = \cdot 0357 \text{ ohm.}$$

$\therefore$ The area of the cable $a = \dfrac{s \times l}{R}$

$$= \frac{\cdot00000066 \times (80 \times 36)}{\cdot0357}$$

$$= 0\cdot0532 \text{ sq. in.,}$$

and from the table the size of cable approximating to this is a $\frac{7}{12}$.

5. A dynamo supplies current to a set of 100 lamps grouped in parallel half-a-mile away.

The feeders have a cross-sectional area of $0\cdot04$ sq. inch, and the resistance of copper is $\cdot00000066$ ohm per inch per square inch. If a current of $0\cdot2$ amp. is taken by each lamp at an E.M.F. of 220 volts, what must be the E.M.F. across the dynamo terminals?

All that is needed in this case is the total resistance of the feeders. Having got this the "drop" along them will be found by the product of current strength and resistance. This drop, added on to the 220 volts will give the E.M.F. required.

$$R = \frac{s \times l}{a},$$

and $s = \cdot00000066$ ohm per inch per half-inch.

$l = 1$ mile $= (5280 \times 12)$ inches.

$a = 0\cdot04$ sq. inch.

$\therefore R = \dfrac{\cdot00000066 \times 5280 \times 12}{\cdot04} = 1\cdot045 \text{ ohms.}$

Total current carried by feeders $= 100 \times 0\cdot2 = 20$ amps.

$\therefore$ Drop along feeders $= 1\cdot045 \times 20 = 20\cdot9$ volts.

$\therefore$ Dynamo must run at an E.M.F. of

$$220 + 20\cdot9 = 240\cdot9 \text{ volts.}$$

6. A dynamo maintains a constant E.M.F. of 110 volts between its brushes. It supplies a power of 10,000 watts to a house 100 yards away. What must be the area of cross-section of the copper cables so that only 4 °/₀ of the power is wasted in them?

Specific resistance of copper is ·00000066 ohm per inch per sq. inch.

The power to be wasted in the cables $= 4\,°/_\circ$ of 10,000

$$= \tfrac{4}{100} \times 10,000 = \textbf{400 watts}.$$

The total current carried in the cables

$$= \frac{10,000}{110} = 90\cdot9 \text{ amps.}$$

The watts wasted $= 400 = \text{current}^2 \times$ Resistance of the cables.

$$\therefore\ 400 = 90\cdot9 \times 90\cdot9 \times R.$$

$$\therefore\ R = \frac{400}{90\cdot9 \times 90\cdot9} = 0\cdot0484 \text{ ohm.}$$

$$\therefore\ \text{area of cross-section} = \frac{\cdot00000066 \times (200 \times 36)}{\cdot0484}$$

$$= \cdot098 \text{ sq. in.}$$

From the tables this cable should be a $\tfrac{19}{14}$ or better a $\tfrac{19}{12}$.

Cable Insulation Regulations. The insulation of cables is a matter of much importance. The insulating material is called the *dielectric*, and it must have a high specific resistance and be capable of withstanding damp.

The Institution of Electrical Engineers' regulations are :

(1) All dielectrics must be damp proof and must have a minimum thickness of 30 mils $(\tfrac{3}{100}$ inch$) + \tfrac{1}{10}$ the diameter of the conductor.

(2) The dielectric must not soften at a temperature lower than 170° F.

(3) The minimum insulation resistance per mile of length shall not be less than that stated in the table, after 24 hours' immersion in water and 1 minute's electrification, at an E.M.F. of 600 volts.

The absolute minimum should be 300 megohms per mile.

Determination of Insulation Resistance of a Cable.
The method of measuring the insulation resistance of a cable

is similar in principle to the method for High Resistance
described in Chapter IV.

A length of the cable is coiled up and placed in a zinc
bath as illustrated in Fig. 73. It is covered with water, and
its ends project as shewn. The resistance is measured between

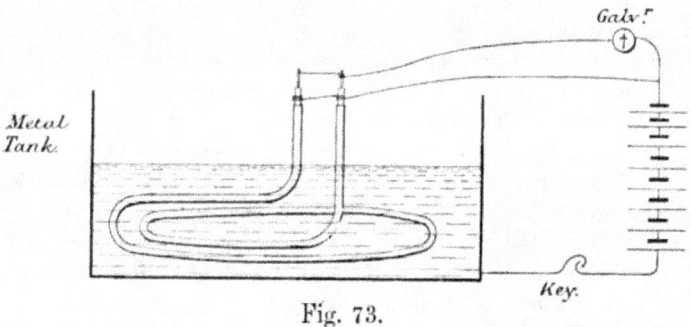

Fig. 73.

the conductor and the bath walls. To prevent errors due to
leakage over the *surface* of the insulating material to the
water and thence to the walls of the tank a device known as
Price's guard wire should be employed. This merely consists
in wrapping a few turns of bare wire around the insulation—
a little distance away from the bare ends of the conductor—
and connecting up as shewn. Any leakage current which
might be set up will not pass through the galvanometer under
these conditions.

The battery *B* must be one of say 500 cells giving a high
E.M.F. of 600 volts and upwards.

Owing to capacity effects the circuit should be "made"
for 1 minute before any reading is taken ; and it is advisable
to short-circuit the galvanometer during this period, since
there may be a rush of current at the first which would tend
to damage the instrument.

The cable should be immersed in the water for 24 hours
before the test is made.

If the galvanometer is shunted the deflexion must be
expressed as equivalent deflexion without shunt : and it is

compared with that produced by a standard megohm under similar conditions.   From this

$$\text{Insulation Resis. of the Cable} = \frac{\text{Resis. of Standard} \times D_S}{D_R},$$

where D_S and D_R are the equivalent deflexions without shunts when the Standard and the Insulation Resistance are in the circuit respectively.

The insulation resistance varies inversely as the length of the cable and should be expressed in *megohms per mile*; but the data of the test should be given, as it is found that the value of the insulation resistance is determined by the E.M.F. of test, the period of electrification and the period of immersion in water.

Hence the necessity of expressing the values according to the definite regulation of 24 hours' immersion, after 1 minute's electrification and at 600 volts E.M.F.

Insulation Grade.   Cable manufacturers have different "grades" of cable, and these grades have different insulation resistance per mile values.   Cable for 100 volt supply need not have such a high insulation resistance as those to be used for 230 volt supply.   The various grades are known as "the 300 megohm (300 Ω) grade" or the "2000 Ω" grade and so on.   This means that the minimum insulation resistance per mile of all cables in that grade is 300 Ω, or 2000 Ω, or whatever the grade may be.   As the big cables are allowed a smaller insulation resistance than the small it follows that all cables of a grade, other than the largest made, will have a *greater* insulation resistance per mile than the standard of the grade.

Wires and Fittings.   With regard to the construction and manufacture of cables and fittings the author can scarcely do better than refer the reader to the various catalogues of the many manufacturers.

Wires and cables are made of copper—and the strands are all of them *tinned*.   This is done to prevent chemical action between the sulphur used in *vulcanising* the rubber insulation and the copper. There is no chemical action between sulphur and tin.

The conductors are covered with layers of pure rubber tape : then with layers of compound rubber of greater mechanical strength than pure rubber, though of less specific resistance. This is then *vulcanised*; a small quantity of sulphur being mixed with the rubber and kept at a high pressure at a temperature of about 270° F. This vulcanising increases the durability and flexibility of the insulation and enables it to withstand higher temperatures.

Then external coverings of compounded waterproof material are put on ; and for special purposes a lead casing or an iron sheathing (armour) is further added.

The main points in connexion with switches are that there shall be no appreciable resistance at the places of contact and no consequent overheating ; the switch should have a long "break" and a spring strong enough to ensure a quick break, so that there shall be no possibility of a permanent arc. To this end small switches must *not* be used to control large currents or even to control small currents in highly inductive circuits. The construction should be such that on switching off it should be impossible to "hold" a switch so that only a small distance (say $\frac{1}{8}''$) separates the places of contact. Bases should be incombustible ; handles should be thoroughly insulated ; covers should be quite clear of internal mechanism, and preferably, should be made of or lined with insulating material. Tests should be made with an E.M.F. 50 °/₀ greater than that to be used, and at the maximum current for which the switch is designed.

With all fittings, high insulation, good connexions, incombustibility, good workmanship, and good mechanical design are essential.

Fuses. Fuses are made either of thin copper or of lead-tin wire, and a fuse of such size is introduced into a circuit so that should the current become greater than that for which it was designed or than any given maximum the fuse will "blow." This is an application of the heating effect. There are thus "3 ampere fuses" which blow at a greater current than 3 amps; and fuses are made up to 500 amperes after this manner. Fuse blocks, or cut-outs, should be of inflammable material, and the fuses themselves should be arranged so that there shall be no chance of their not breaking the circuit completely when they blow. The usual length of a fuse is about 3 inches, in order to assist in this complete break. Hard metal fuses are preferable to those of soft metal.

Electro-magnetic cut-outs are used as "main fuses" on large current supplies, the heating fuse being objectionable and dangerous for obvious reasons when it has to be made for large currents.

Regulations for Insulation Resistance of an Installation. When an electric installation is completed it is necessary that an *insulation test* of the entire system shall be taken. This is insisted upon by the electric supply company and by the insurance company interested.

The insulation resistance of an installation is often defined as the *megohms per lamp*—but it must be understood that the total insulation resistance will be the megohms per lamp divided by the number of lamps. For this the number of lamps should be reckoned according to equal c.p. (e.g. 8 c.p.). Thus a 32 c.p. would be equivalent to four lamps on an 8 c.p. per lamp basis.

Different companies have different limiting values for the insulation resistance of an installation.

The Institution of Electrical Engineers regulation is that the minimum insulation resistance shall be 10 megohms divided by the maximum current supplied.

The leading Insurance Companies state that the leakage $\left(\text{viz. } \dfrac{\text{E.M.F. of supply}}{\text{Insulation Resistance}} \right)$ shall not exceed $\dfrac{1}{20,000}$ of the maximum current.

Electric supply companies generally state the megohms

per lamp—and these vary from 10 to 80. The different standards of insulation will depend chiefly upon the E.M.F. of supply.

Testing an Installation. The insulation resistance of an installation is measured by having all lamps, fuses and switches *on* and measuring the resistance between the main conductors and the earth—i.e. between the bus bars on the main switchboard and the earth. It should be measured at double the E.M.F. for which the system is designed, and after not less than one minute's electrification.

Another test is to remove all the lamps from their holders and with switches and fuses *on* to measure the insulation between the two main bus bars.

Special " testing sets " have been designed for measuring these insulation resistances. There are many different forms, some being direct reading and others being a portable arrangement for measurement on the principle described on page 72.

In making a test, if the whole insulation should come below the required minimum, each section should be tested separately in order to find out where the faulty part lies. When the section has been found then the branches of that section should be tested—and ultimately the fault may be tracked down and then remedied.

In new buildings the insulation resistance is often found to be below regulation. But after some days, during which fires are burned in the building, it will probably be found to comply with the necessary conditions.

CHAPTER VIII.

MAGNETISM.

THE reader is already familiar with the general elementary phenomena of magnetism treated from a purely qualitative standpoint. The application of these phenomena to the subject of electrical engineering is of supreme importance, for in many applications of electricity for practical purposes it is the magnetic effect of a current which plays the principal and all-important part. This has been seen already to some extent in the construction of measuring instruments, galvanometers and self-regulating arc-lamps, etc.; but these are of small importance compared with the application of magnetism to motors, dynamos, alternators, and transformers. Before proceeding to discuss these applications of electricity, it is therefore necessary to extend the knowledge of magnetism so far assumed, and to put that knowledge on a quantitative basis.

Fundamental phenomena. A magnet can attract iron filings to itself: it can exert certain effects, called magnetic, in the space about it; if it be suspended so that it can move in a horizontal plane about a vertical axis, it will come to rest in a definite position—a position which is determined by the effects of other magnets in its neighbourhood, or by the effects of the earth's magnetic properties. If it is affected only by the earth's magnetic field it will come to rest with one end pointing towards the North, and the other towards the South. These ends will always point in the same relative direction, and they are called the poles of the magnet. Since they point definitely and always in one direction, they appear to differ in magnetic properties. They are called by different names therefore. The pole pointing to the North is called the *N*-seeking or the marked pole. The other is called the *S*-seeking or the unmarked pole. Poles with similar tendencies repel each other : poles with opposite tendencies attract each other.

If a small pivoted magnet (a compass needle) be placed in the neighbourhood of a large magnet, it is affected by it. Its tendency to point North and South is there just the same, but the magnetic forces exerted by the magnet cause a tendency to point to some other direction. What this direction will be depends upon the position of the needle with respect to the magnet ; upon the position of the magnet with respect to the earth's magnetic forces ; and upon the relative magnitudes of two tendencies caused by the magnet's and the earth's magnetic forces respectively.

The magnetic field.   The space about a magnet within which magnetic effects, such as the foregoing, are apparent is called the *magnetic field* of the magnet. It is perhaps a loose definition, because the limitations of a magnetic field are determined apparently by the sensitiveness of the arrangements used for observing these effects. However, it is all that is required, for if no effects are apparent then the particular apparatus in use is outside the magnetic field for all practical purposes. Of course another piece of apparatus might be *in* it under equal conditions, and the fact would have to be recognised. The point is that the definition does not intend for a moment to put any quantitative limitations to the magnitude of the space called the magnetic field. There is no desire to say, for example, that the field only exists within a spherical space of three feet radius from the centre of the magnet. It may exist throughout all space—or it may exist only within a very limited radius—according to the means employed for detecting it.

Unit magnetic pole and unit magnetic field. This magnetic field then is a space within which magnetic forces are exerted. Now all forces have magnitude and direction; and the effect of a force can only be determined when its direction and magnitude together with its point of application are known. Hence it is necessary to have some method of measuring magnetic forces.

To this end some unit of magnetic force must be defined; *and a magnetic field of unit strength is said to be that uniform field which acts with a unit of force* (1 dyne) *upon a unit magnetic pole.*

This definition involves another, namely the significance of a unit magnetic pole, before it can be extended.

It is impossible to obtain a magnet having only one pole. That fact is the basis of the molecular theory of magnetism. But it has been found convenient to *assume* that such a thing can exist. This assumption—in itself an assumption which facts cannot justify—simplifies the fundamental definitions of units, and does not introduce any fallacies into the measurements which are based upon it.

Now a single pole would exert an attractive or a repulsive force upon another single pole, and this effect would depend upon the *strength* of the poles and upon their distance apart, and upon the nature of the medium separating them. A magnetic pole *of unit strength* is defined conventionally as *that pole which attracts or repels another pole of equal strength with a force of 1 dyne when they are 1 centimetre apart in air.* The poles are supposed to be concentrated at points, and it is when these points are one centimetre apart that the conditions of the definition are fulfilled.

The reader is referred to the quantitative law of magnetism known as Coulomb's law, that the force between two magnet poles varies directly as the product of their strengths and inversely as the square of the distance between them, assuming a uniform and constant medium of separation. Therefore it may be said that

Force between two poles varies as

$$\frac{\text{strength of one} \times \text{strength of the other}}{(\text{Distance between them})^2},$$

or
$$F \propto \frac{m \times m'}{r^2},$$

where F is the force, m and m' the strengths of the poles, and r the distance between them.

If m and m' be expressed in terms of the unit pole and if r be measured in centimetres, it follows that the force between them must be expressed in dynes, and that therefore the sign of equality may be substituted for that of variation under these conditions.   Thus

$$F = \frac{mm'}{r^2} \text{ dynes, in air,}$$

and the poles m and m' will be of unit strength when the force F is one dyne when r is one centimetre.

Force at a point in a magnetic field.   Now if a single pole were placed in any magnetic field one can easily understand that magnetic forces would act upon it.   Let it be assumed that a magnet pole $+ m$ has been placed at the point P in the neighbourhood of the magnet QR, as shewn on Fig. 74.   Between the pole $+ m$ and the pole $+ Q$ of the magnet there will be repulsion along QP.   Between $+ m$ and the pole $- R$ there will be attraction along PR.   Clearly, the pole will not be constrained to move in either of these directions, but in some sort of intermediate direction which will be determined by the relative magnitudes of the forces of attraction and repulsion, and the angle between their directions.

But the actual direction of the *resultant* of these forces at this point P may be readily determined, for

$$\text{the force along } QP = \frac{mM}{QP^2},$$

$$\text{and the force along } PR = \frac{mM'}{PR^2},$$

where M and M' denote the strength of poles Q and R respectively. Now M and M' belonging to one magnet must

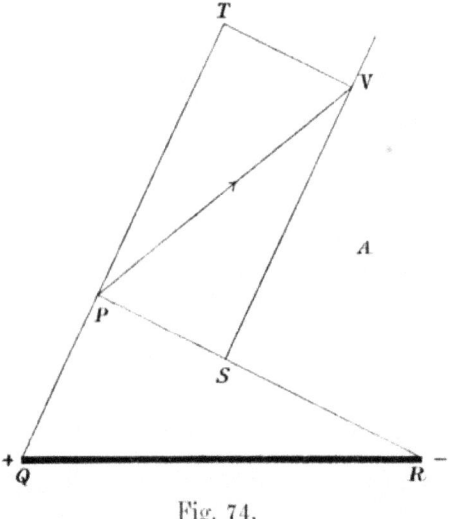

Fig. 74.

be equal. Therefore it is seen that the numerators of these expressions are equal, and thus

$$\frac{\text{Force along } QP}{\text{Force along } PR} = \frac{\dfrac{mM}{QP^2}}{\dfrac{mM}{PR^2}} = \frac{PR^2}{QP^2}.$$

That is to say, the forces along QP and PR are proportional to the squares of PR and QP respectively—inversely proportional, in short, to the squares of the distances.

Thus the direction of the resultant force at P can be simply determined by the principle of the parallelogram of forces. Along PR cut off a portion PS proportional to QP^2. Along QP produced cut off a portion PT proportional, on the same scale, to PR^2. Complete the parallelogram about PT and PS by drawing TV parallel to PS, and SV parallel to *PT*. Then *PV*, the diagonal, will represent the *direction of the resultant* of the forces along *QP* and *PR, at the point P.*

But having determined this there is only one fact known, that the direction of the resultant, that is the *actual,* magnetic force at the point *P* is in the direction of *PV*. At the points *B*, or *A*, or *V*, the direction of the actual magnetic forces due to the magnet must be determined independently, and indeed this applies to every point in the magnetic field of the magnet.

Lines of force. There is a very convenient and simple means of representing graphically the direction of a magnetic field at all points by means of lines known as *lines of force*. The reader is more or less familiar with the methods employed for the delineation of these lines, and he should now endeavour to appreciate their precise significance and their limitations.

If a single *N*-seeking pole could be procured, and if it were capable of motion in any direction, then if placed at any point, say *P* on Fig. 75, it would be acted upon by definite

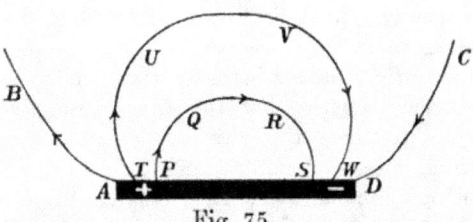

Fig. 75.

forces. The marked (+) pole of the magnet would repel it, and the unmarked (−) pole would attract it, but the actual direction of its motion at the point *P* would be determined by the resultant of these forces. It would be found ultimately to proceed along a path such as *PQRS*, for at every new position different forces (in magnitude and direction) would act upon it and there would be a continually changing resultant force, both in magnitude and direction. If it be started at T, it would be found to follow a path $TUVW$; if started at A, it would follow $AB-CD$. These paths may be drawn by careful application of the foregoing laws—they are not mere approximations—and ultimately a sufficient number may be drawn to represent the actual direction of the magnetic field of the magnet at every point, using the word point in the Euclidean sense.

These lines are called lines of force, and they represent *directions*—and *only* directions—of magnetic forces. They enable one to form a mental picture of a magnetic field, especially when they are ascribed certain properties. Michael Faraday, that great experimentalist and master of mental conceptions of phenomena, *assumed* that these lines had a physical existence; that they were like stretched strings ever trying to get shorter; that they repelled one another; that, in short, there was tension along them and pressure across them. If the idea be accepted, one gets a mental conception of the physical nature of a magnetic field.

But when that has been gained, the materialistic ideas involved may be gradually discarded. The lines of force have no physical existence; they are but lines drawn on paper, or drawn mentally to indicate direction. But the ideas of tension along these directions can remain; the tension takes place in a direction, and *not* in an existing line like a stretched string. It may be acting just the same. And the pressure across these directions—that idea may remain also dissociated from material lines.

The mapping of magnetic fields is a most important part of the laboratory work of the beginner in magnetism; and it is assumed that the reader is so far familiar with this that he is able to draw a map of a magnetic field due to different combinations of magnets, from his own mental conception, without the aid of iron filings or compass needles.

Thus on the general subject of magnetic fields it is sufficient to say that a magnetic line of force is that line which would be described by a single N-seeking magnet pole free to move in any direction ; and that the direction of a line of force. is the direction in which this pole would move.

Field strength defined in terms of a number of lines of force. To return to the definition of the strength of a magnetic field it is found that the strength of a field is determined by the *force* (actual) with which the field acts upon a unit magnetic pole. But the strength of a field is not a constant quantity at all points. Hence, except in special cases where the uniformity of the field is stated, one can only speak of the strength of the magnetic field *at a point*.

There is another convention adopted for the definition of the strength of the magnetic field at a point. According to this *the strength of the magnetic field at a point is measured by the number of lines of force which cut a square centimetre at that point in a plane at right angles to their direction.*

Thus if a number of lines of force be imagined as cutting this paper at right angles—either coming up or going down through it—then the number which cut a square centimetre represents the strength of the magnetic field at the place where they were counted.

But it must be added hastily that since lines of force have not physical existence they are incapable of being counted. However, the idea is helpful for many purposes, and another convention has been adopted that they may be counted.

A magnetic field of unit strength acts with a force of one dyne on a unit pole. This field is said to have *one line of force per square centimetre*, the measurement being made at right angles to the direction of the lines.

Thus there are two modes of expressing the strength of a magnetic field at a point, one by the force in dynes which is exerted upon a unit pole at that point, and the other by the number of lines of force per square centimetre at that point, at right angles to their direction.

These expressions are *numerically equal*; they amount to one and the same thing ultimately. The number which expresses the magnitude of the force in dynes on a unit pole at a point, also expresses the number of the lines of force (according to the convention) which cut a square centimetre at that point. Either may be used, but two have been adopted for the ultimate object of simplicity.

The number of lines of force emanating from a unit pole. As an application of these ideas it might be well to determine the number of lines of force which proceed from a unit pole.

If a pole of m units be imagined at the centre of a sphere of radius r centimetres then the force due to that pole at any point on the surface of the sphere would be

$$= \frac{m \times 1}{r^2} \text{ dynes.}$$

This represents the force which would be exerted on a unit pole by the pole m, if the unit pole were placed at the surface of the sphere.

Therefore according to the convention the strength of the field at the surface of the sphere is $\frac{m}{r^2}$ units.

Therefore also there must be $\frac{m}{r^2}$ lines of force cutting each square centimetre of the sphere.

Now the area of the surface of a sphere is $4\pi r^2$ square centimetres where r is the radius in centimetres and π the relationship between circumference and diameter and equal to 3·142.

Thus the total number of lines of force cutting the sphere must be

$$4\pi r^2 \times \frac{m}{r^2} = 4\pi m \text{ lines.}$$

Hence it follows that $4\pi m$ lines proceed from a magnet pole whose strength is m units. Therefore the *number of lines proceeding from a unit pole must be 4π lines.*

The foregoing definitions and laws have all been stated for single poles. The fact remains that single poles cannot exist, but it can be shewn that the laws are correct. In applying them one has merely to consider all the forces acting at a point and to determine the resultant of them. This may be done graphically by application of the methods employed in the subject of Mechanics for the determination of the resultant of many forces. The reader would be well advised to attempt some graphical solutions for himself. He should determine the magnitude and direction of the resultant force at some point in the neighbourhood of a magnet. He can assume that the strength of each magnet pole is 50 units, say; and the length of the magnet is 30 centimetres. He can thus obtain absolutely definite results.

And further, if he keeps the pole strengths constant and varies the length of the magnet, he will find out what an important matter that length is.

Moment of a magnet. In this connexion it should be explained that there is a great distinction between the strength of a magnet *pole* and the *strength of a magnet*. The strength of a magnet pole may be measured by the number of lines of force which leave it or enter it according to its polarity. (It will be seen later how the strength may be determined in these terms.) Now the lines of force which enter one pole must stream through the magnet and leave the other pole. The poles of one magnet are thus equal in strength. But two magnets may be obtained having the same *pole strengths*, yet if they be of different length it will be found they are not capable of producing equal effects in the space about them. That is to say they are not of equal strength as a whole, even though they have equal pole strengths. The longer magnet will be capable of producing greater effects than the shorter, and this may be readily proved by graphical methods or may be easily reasoned out. If a magnet of constant pole strength be gradually shortened, its poles must be brought nearer and nearer to one another. Thus the forces (+ and −) exerted by them become more and more nearly in one straight line and more and more nearly equal and opposite in magnitude

and direction. And ultimately if the magnet be imagined to
have no length—the two poles merged into the same point—
it can be seen that the resultant force at all points about the
magnet must be zero, for the forces due to the poles must be
equal and opposite at every place.

And on the other hand as the same magnet is mentally
increased a limit will be reached when the poles are so far
apart that they do not produce any measurable effect on one
another ; and under such conditions the force at a point can
be regarded as being due only to the nearer pole of this long
magnet.

The strength of magnets then as a whole, comparing the
strengths by the forces which these magnets are respectively
capable of exerting at a certain distance in some definite
direction from their centres depends upon the pole strength
and upon the lengths of the magnets—or more strictly upon
the distances between the two poles. The expression the
moment of a magnet is used to convey the idea of the strength
of a magnet, and the magnetic moment is measured by the
product of the strength of one of its poles and the distance
between them. It is generally denoted by M.

Thus $$M = 2l \times m,$$

where $2l$ is the distance between its poles and m the strength
of one of the poles. The general expression is given, and $2l$
is used to denote the distance between the poles for arith-
metical reasons. The expression the *moment of a magnet* is
not unlike the mechanical expression " moment of a couple."
For if a magnet be suspended so as to move in a horizontal
plane about a vertical axis at its centre, then its tendency to
set itself in the magnetic meridian would be determined by
its pole strength and the distance of each pole from the
turning point, namely the centre of the magnet. Thus if l
denotes the distance of each pole from the centre and m the
strength of each pole, then the tendency of the magnet to *set*
itself would be measured by the product of the force exerted at
either end and the distance between them when the magnet is

13—2

placed E. and W.   The actual force at the pole would depend upon the pole strength and upon the strength of the field in which it was swinging.   But the tendency will be *proportional* to $2lm$, which is called the moment (M) of the magnet.

Molecular Theory of Magnetisation of Iron.   When a piece of unmagnetised iron is brought into a magnetic field it becomes magnetised.   When it is removed again it returns almost to its original state.   The phenomenon is called magnetic induction, and it should be borne in mind that *all* processes of magnetisation are merely inductive.

Now on magnetisation of a piece of iron there is neither loss nor gain in weight ; there is absolutely no evidence of anything having been given to or taken from it ; the strength of the magnet used for the purpose is not altered by the process, and thus one is brought to the conclusion that magnetisation might be some quality of a body like hardness or softness which can be produced in varying degrees by alteration of the physical structure of the body.   The process of reasoning thus far is not so self-evident and straightforward as the above would seem to indicate.   It is rather in the light of experimental facts that the conclusions are arrived at, but the reader should be aware of the elementary phenomena and be familiar with the rudiments of the molecular theory of magnetism.

According to this conception every molecule of a piece of iron is always a complete and separate magnet, and when magnetisation is attained the molecules arrange themselves in the direction of the magnetising field, their marked poles all pointing in one direction.   This is the conclusion which is strengthened by the fact that when a magnet is broken there are always as many complete magnets as there are pieces ; that at each place of breaking a free marked and a free unmarked pole is produced ; and that the original magnetic state of the magnet can only be reproduced by piecing it together again with opposite poles in contiguity.

In the unmagnetised state the molecules are assumed to

arrange themselves in closed magnetic circuits, leaving no free poles and shewing no external signs of magnetisation.

This theory is based on strong experimental facts. A "tick" is heard on magnetisation of a piece of iron and heard again when the magnetising forces are withdrawn. The same "tick" in an amplified and extended form is heard in the humming of an alternate-current transformer. The tick is the noise produced by the molecules changing their positions. The hum of the transformer is a musical note which corresponds in pitch to twice the number of alternations of the current per second. Every alternation means a reversed magnetic field, and consequently reversed magnetisation of the iron. According to the theory every reversal of magnetisation means a change of position of the molecules. This is heard, and so rapidly are the successive "ticks" that a musical note is produced.

Further, the iron becomes heated by the process of magnetisation. If magnetisation is the result of molecular motion then the heating is accounted for, since matter in motion always meets with friction, and as a result the energy required to overcome it is changed into heat.

In magnetisation iron alters in length, but not in volume.

Magnetisation is assisted by tapping the iron when the magnetising forces are acting. Conversely de-magnetisation is assisted by tapping when the magnetising forces are removed. It may be reasoned that magnetisation is a state of strain, and that a piece of iron is ever trying to return to its original state.

Again, energy must be expended to magnetise iron. If it be magnetised by rubbing with a magnet the magnet is not altered in strength. Where does the energy spring from? From the individual who is doing the rubbing. More work has to be done in pulling the magnet along the iron than is required to pull it along some non-magnetisable body under equal conditions. According to the theory the energy is being used in breaking up the closed circuits of the molecules and in forcing them into the line of magnetisation.

Energy of a Magnet. It would seem then that the energy of a magnet is merely potential energy—energy of state—energy of position of its molecules. It will give up some of this energy when some of its molecules go back to their original conditions. *Only* thus can it give out energy. That is to say, a magnet cannot do *work* unless it gives up some of its energy and loses some of its magnetisation. The reader may think of the statement that one magnet may be used to magnetise any number of pieces of iron and yet not lose any of its own magnetic properties. True! But it doesn't do the work—it is only the *means* of exerting the *force*—it is equivalent to the string by which one pulls an object—the work is done by the individual who overcomes the resistance offered and moves the point of application of the force. And it is for this purpose that a magnet is of use, as a means of exerting force and not as a source of energy in itself. It is a metaphorical hook which catches up a metaphorical eye. This is a matter the full appreciation of which is of no small importance.

Degrees of Magnetisation: Magnetisation Curve. Since the process of magnetisation consists in arrangement of molecules it would seem that a limit must be reached when all the molecules of a piece of iron have been set. This again is borne out by experimental fact. There is a limit to the degree of magnetisation of every piece of iron and when the limit has been reached the iron is said to be *saturated*, a word which is a relic of the old-time fluid theories of magnetism.

This may be demonstrated by an exceedingly useful experiment. A reflecting *magnetometer* is set up like a galvanometer with a lamp and scale. This magnetometer merely consists of a small disc of aluminium (say half an inch in diameter), which is suspended by a piece of fine, unspun silk. On the back of this disc a piece of magnetised watch spring is fastened, and on the front is affixed a small mirror. This arrangement is boxed in to be free from air draughts, and sometimes the "needle" swings in a copper tube, which assists in damping the vibrations by the effects of induced eddy currents.

This magnetometer is set up so that the needle can move freely and the spot of light on the scale is adjusted to the zero position. The needle will be controlled by the earth's magnetic field. Behind the magnetometer a solenoid AB, shewn on Fig. 76, is arranged in a vertical position. This should

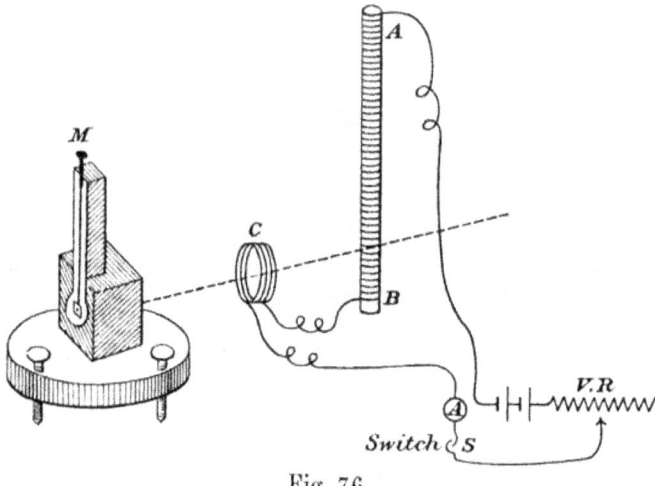

Fig. 76.

be placed so that the centre of its cross-section is on a line which passes through the centre of the magnetometer needle at right angles to its axis when it is in its normal condition of rest. In series with the solenoid is a coil C, called a *compensating coil*, and this is placed between the solenoid and the magnetometer as shewn. These are connected to a battery, a resistance which may be easily and gradually varied such as a wheatstone rheostat, an ammeter or other current indicator, and a switch.

When a current is passed through the solenoid its magnetic effect will cause the magnetometer needle to be deflected.

However, this is not desired, for the solenoid is to be used as means of magnetising the piece of iron which is to be experimented upon. The compensating coil is therefore adjusted so that its magnetic effect compensates for that of the solenoid at the magnetometer needle, and this result is obtained when the spot of light returns to its original position of zero. Now as the solenoid and compensating coil are in series this balance will hold for all currents through them, but it is advisable to adjust when the maximum current to be used is flowing through the solenoid.

Having made these preliminary adjustments the current is switched off, and the resistance made as large as is necessary. The piece of iron to be tested should be in the form of a long, thin bar—a piece of iron wire will serve admirably. This should be rather shorter than the length of the solenoid, and it should be put into the solenoid so that its lower end is in the same horizontal plane as the magnetometer needle, and fixed with a cork top and bottom. The iron and the solenoid being long, the effect of the upper ends may be neglected.

Now when the iron is first put in, in an unmagnetised state one would not expect any deflexion of the needle to take place, since there is no current in the solenoid. However, it will be found that the needle does deflect, shewing that the iron is slightly magnetised. This will be produced by the vertical component of the earth's magnetic field, and it may be compensated by winding some turns of wire on the outside of the solenoid and passing a constant current through it in such a direction and of such strength that the earth's effect is neutralised and that the spot of light once more returns to zero.

The preliminaries are now completed, and the whole arrangement provides a means of applying a magnetising force of gradually increasing magnitude and of measuring, in some terms or other, the corresponding degrees of magnetisation produced. The magnetising force is the magnetic field inside the solenoid which is proportional to the strength of the current passing. Thus the readings of the ammeter will give a measure of magnetising force.

The magnetometer needle will be deflected by the iron as it is magnetised, and *only* by the iron, since other effects have been compensated. The tangent of the angle of deflexion will be proportional to deflecting force, which in turn will be proportional to the degree of magnetisation of the iron. Since the deflexions are obtained with a spot of light, like those of the mirror galvanometers, they are proportional to the tangents of the several angles of deflexion of the needle. Hence the deflexions along the scale give a measure of the degrees of magnetisation.

A small current is passed through the solenoid. This magnetises the iron wire slightly. The current strength and the corresponding deflexion of the needle are noted. The resistance is reduced a little, and the new current and corresponding deflexion are noted. And so the experiment is continued, gradually and carefully, until a stage is reached when the deflexion does not appreciably increase for comparatively large increments in the current strength.

To examine the results of the experiment a curve should be plotted with current strength as abscissae and deflexions as ordinates. Fig. 77 is such a curve of results.

It will be noted that this curve may be divided up into three sections, namely *OE*, *EC*, and *CF*.

The Three stages of Magnetisation. These sections have different characteristics and represent three stages in the process of magnetisation.

In the first, the degree of magnetisation is fairly proportional to the magnetising force ; and if the magnetising force were removed the iron would be found to return to its original state. This stage is known as the *elastic stage*.

At the end of this stage it is found that the small increase in the magnetising force produces a large increase in the degree of magnetisation, so that the curve suddenly becomes steep, as is shewn by the part *EC*. This stage is known as the *catastrophic stage.* Once the iron has reached this stage, a removal of the magnetising force will *not* restore it to its original condition. It will retain some of its magnetisation.

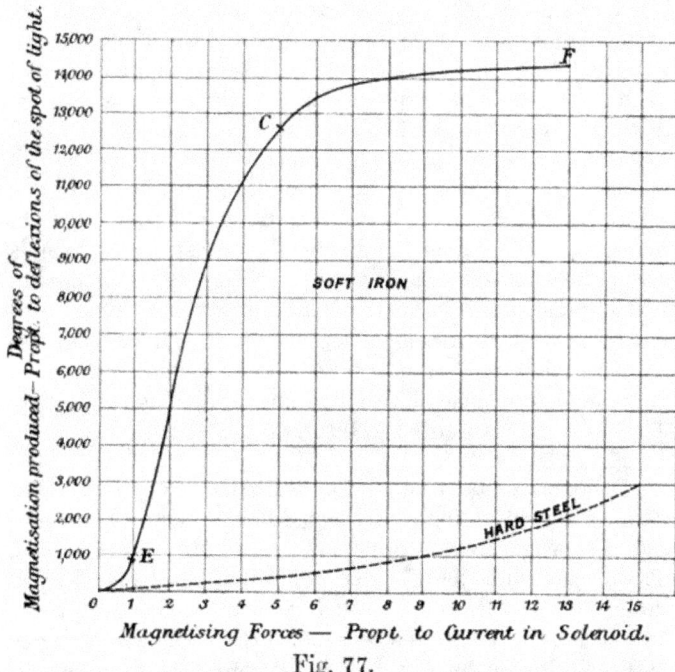

Fig. 77.

The third stage, *CF*, shews that a large increase in the magnetising force produces but a small increase in the degree of magnetisation, and finally a limit to the latter is reached, notwithstanding the fact that the magnetising force is being increased. This is the *final stage*.

These stages admit of explanation on the molecular theory, and they can be clearly demonstrated by a model known as Ewing's model, which consists of a group of small pivoted needles, some 50 or more, arranged on a rectangular plate. These are supposed to represent the molecular magnets of a piece of iron, and they will be found to arrange themselves finally in approximately closed magnet circuits when they are unaffected by any magnetic field. This condition will represent the unmagnetised state.

A magnetic field can be caused to act upon them, by arranging an open coil so that it surrounds them and passing a current through it. If the current be gradually increased as in the foregoing experiment, the three stages can be distinctly seen. In the first, a *few* of the needles are acted upon, they make a half-hearted attempt to set along the line of magnetisation; but if the current be switched off they return back in apparent haste to form their old circuit. On increasing the current gradually the catastrophic stage is shewn in unmistakeable manner. Nearly all the needles swing round into the line of magnetisation. A few will remain, but on increasing the current still more they too will eventually come into line, and the final stage is reached. No more can be done since all the molecules are set in line. The system is magnetised to saturation.

Difference between Magnetisation of Iron and Steel. On subjecting samples of different irons to the process of magnetisation different results will be obtained. These results are found to be dependent upon the physical differences of the samples, such as hardness and softness, and density. Hard iron will not attain the same degree of magnetisation as soft iron under precisely equal conditions. And the stages of magnetisation will not be so clearly defined.

The distinctions are shewn to the utmost by the comparison of soft iron and very hard steel. Fig. 78 is a magnetisation curve of a sample of steel obtained under equal conditions to Fig. 77, and drawn on the same scale. The two may be directly compared, and it is seen firstly that the degree of magnetisation of the steel for a given magnetising force is always less than for the iron. Secondly, the stages of magnetisation are less clearly defined; the catastrophic stage is more gradual, and so is the final. The curve more nearly approaches to a straight line ; the degree· of magnetisation is more nearly proportional to the magnetising force up to saturation limit.

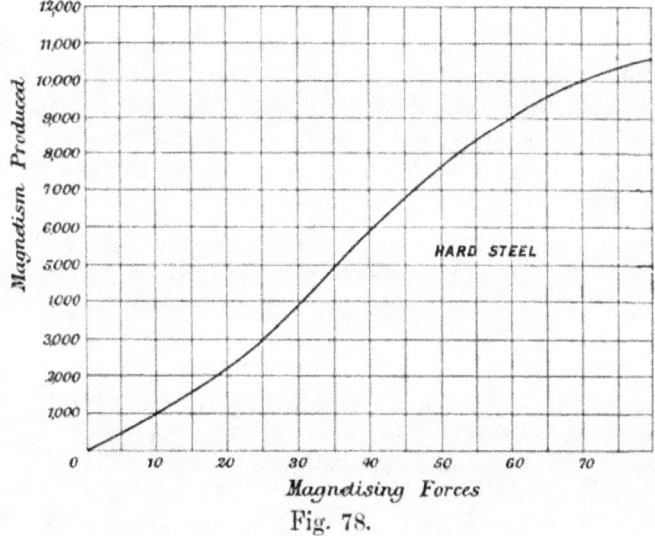

Fig. 78.

Permeability. The foregoing discussion shews that there is some kind of a relationship between the degree of magnetisation attained and the magnetising force acting. It also shews that this relationship is not a constant quantity or the curves of magnetisation would have been straight lines.

The relationship is termed the *magnetic permeability* of the substance and it may be defined in two ways. In the case of a magnetisable substance it may be defined as *the degree of magnetisation attained per unit of magnetising force applied.* Or more generally it may be defined as the *conductivity of a substance for magnetic forces compared with air.*

All known substances allow magnetic forces to act through them. The influence of a magnet can be exerted through glass, wood, or metal; through anything, or as far as can be determined through empty space. But different things transmit the magnetic forces differently; they may be said to have different conducting-properties. The expression *magnetic permeability* is used to express the conductivity of a substance for magnetic forces as compared with air under equal conditions.

Now magnetic forces may be expressed in terms of the number of lines of force per square centimetre, measured at right angles to their direction. Thus the magnetic permeability (denoted by μ) of a substance may be expressed as

$$\mu = \frac{\text{number of lines per square centimetre in the substance}}{\text{number of lines per square centimetre in air under equal conditions}}.$$

If B denote the number of lines of force per square centimetre in the substance; and H the number of lines per square centimetre in the air under equal conditions, then

$$\mu = \frac{B}{H}.$$

The lines denoted by B are generally termed *lines of induction*, so that B denotes the *induction* per square centimetre produced by some magnetising force H.

––––––

The magnetisation curves, Figs. 77 and 78, of iron and steel, really depict the relationship between B and H. B is the degree of magnetisation and H is the magnetising force: but it must be remembered that B is used to represent the degree of magnetisation per square centimetre of cross-sectional area of the substance, and H the magnetising force per square centimetre. The *total induction*—that is the total number of lines of induction through the substance will clearly be represented by

$$N = B \times a,$$

where N is the total induction and a the area of cross-section of the substance.

If a long iron rod of area of cross-section a square centimetres be introduced into an air space, where the magnetic force is H lines per square centimetre, it will become magnetised.   It may be supposed to acquire a pole strength of m units.

Previous to the introduction of the iron there were in the air space now occupied by it $H \times a$ lines of force.

But the iron has acquired a pole strength m and therefore there are $4\pi m$ lines leaving it or entering it as the case may be.   Hence in the iron there are now

$$(H \times a) + 4\pi m \text{ lines of force.}$$

Therefore $$B \times a = (H \times a) + 4\pi m,$$

$$\therefore \ B = \frac{(H \times a) + 4\pi m}{a},$$

$$B = H + \frac{4\pi m}{a}.$$

This indicates a mode of measuring B by measuring the pole strength of the sample under test and the area of cross-section.

Intensity of Magnetisation.   The *intensity of magnetisation* (denoted by I) is a quantity which denotes the magnetic moment M of a magnetised sample per unit of its volume.   It is apt to be mistaken for the total induction or for the magnetic moment of the whole.   Now when a piece of iron has been magnetised the total induction will be equal at all parts in its cross-section.   Again, if the sample be cut transversely into two or more strips the total induction of each strip will be proportional to their cross-sectional areas. Thus it can be seen that if the sample be cut up in any way into pieces the magnetic moments of each piece will be proportional to its volume; and the magnetic moment of each piece will be the same fraction of the magnetic moment of the whole, as its volume is a fraction of the whole volume

$$\frac{\text{magnetic moment of piece}}{\text{magnetic moment of whole}} = \frac{\text{volume of piece}}{\text{volume of whole}}.$$

Thus the *moment per cubic centimetre* expresses a definite magnetic state of both the whole bar and any part thereof; and it is called the *intensity of magnetisation.*

The relation of I to the pole strength of a magnet is a simple one. The pole strength is, say, m; the length is $2l$ centimetres and the area of cross-section a square centimetres. Therefore the moment of the magnet $M = 2ml$ and the volume of the magnet $= 2la$.

$$\therefore \ I = \frac{\text{moment}}{\text{volume}} = \frac{2ml}{2la} = \frac{m}{a}.$$

Referring back to the expression on page 206, that

$$B = H + \frac{4\pi m}{a},$$

it is seen that this may be written

$$B = H + 4\pi I.$$

Magnetic Susceptibility. There is another expression which refers to the "readiness" for magnetisation of a substance. This does not coincide exactly with the term *permeability*, but conveys similar ideas in a more restricted sense. This is *magnetic susceptibility*, and it denotes the intensity of magnetisation produced when a unit of magnetising force is acting.

Thus the magnetic susceptibility

$$K = \frac{\text{intensity of magnetisation}}{\text{magnetising force}}.$$

$$\therefore \ K = \frac{I}{H}.$$

It has been seen that $B = H + 4\pi I$.

Now $\qquad\qquad K = \frac{I}{H}, \quad \therefore \ I = KH.$

Hence $\qquad\qquad B = H + 4\pi KH,$

$$\therefore \ B = H(1 + 4\pi K),$$

and since $\qquad\qquad B = \mu H,$

$$\therefore \ \mu = (1 + 4\pi K).$$

Variation of μ **with** H. The permeability of a sample of iron or steel varies with the magnetising force. This is unfortunate for practical purposes since one cannot state definitely the μ of a given sample. One is bound to give its variations. Fig. 79 is a magnetisation curve shewing the

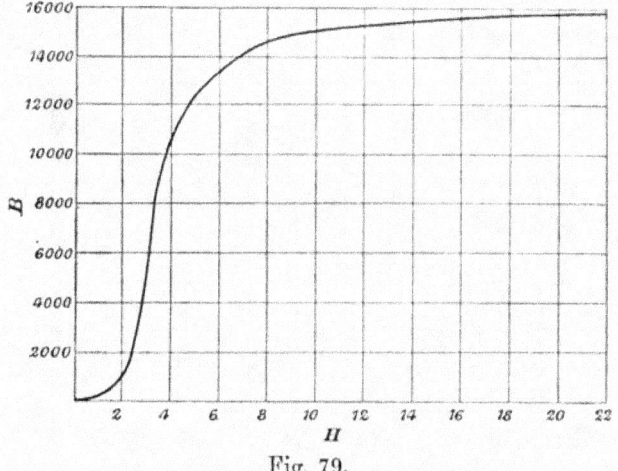

Fig. 79.

true values of B as ordinates and of H as abscissae. For any value of H the permeability is B/H where B is the value shewn on the curve.

Fig. 80 is the *permeability* curve of the same sample of iron, shewing the relationship between μ and H. The ordinates shew μ (viz. B/H) and the abscissae shew H. It is seen that the permeability is a maximum when H is 4·5, and that it decreases for further increases in H. These two curves should be carefully compared by the reader. It can be seen too that if this iron were to be used to produce a magnetic field—say in an electromagnet—it would be wasteful to apply a greater H than 4·5 units. This is an important practical point.

It will be seen that the permeability of a sample of iron or steel will depend upon its hardness or softness. Generally speaking it may be said that soft iron is more permeable than steel, and that the permeability decreases as the hardness increases.

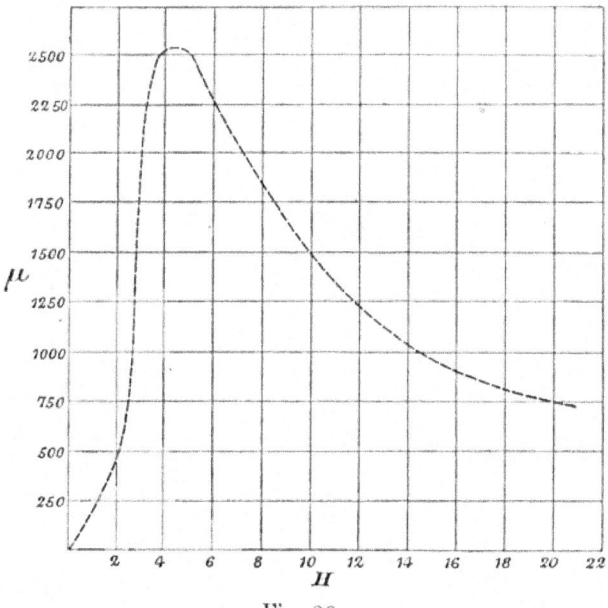

Fig. 80.

Hysteresis. All samples of iron or steel will be found to retain some degree of magnetisation after all magnetising forces are removed. Indeed there is always a tendency for the substance to retain its magnetic condition when an attempt is made to alter it. The molecular magnets would appear to resent any alteration of their positions, and this resentment seems to be greatest when the hardness of the substance is greatest. This tendency is most clearly shewn when a substance is subjected to a cyclic change of magnetisation—that is to say when it is magnetised to saturation in one direction and then in the other direction and back again to the first.

P. Y.   14

This may be done by means of the apparatus described·for the delineation of a magnetisation curve (page 199). The only addition to be made is a reversing key by which the magnetising current may be reversed without alteration of wires.

When all the compensations have been made, a magnetised sample of, say, soft iron is put into the solenoid and the magnetisation data are taken—deflexions representing B and currents representing H.

When the iron has been magnetised to saturation, the cyclic process can be commenced. The current strength is gradually decreased and readings for B and H taken. It will be noted immediately that these do not correspond to those taken on the magnetisation for equal values of H. When the current, and therefore H, has been reduced to zero, it will be found that there is a large deflexion representing B. This represents the *residual* effect.

Then the current, and therefore the direction of the magnetising force H, is reversed, and slowly increased. The effect of this reversed force will be to diminish the reflexion and a point will be reached when this is zero. At this point therefore the magnetisation produced, that is B, is *nil*, but to bring about this effect a negative magnetising force is necessary. (If this be removed B will increase again, so that the de-magnetisation is not permanent or natural, but temporary and forced.)

The magnitude of the reversed H necessary to reduce B to zero represents the *retentivity* or *coercive force* of the sample under test.

The reversed H is gradually increased further, and then the magnetisation B will also become reversed and rapidly reach the final stage of reversed magnetisation. Reversed saturation will occur at an equal $-H$ to the $+H$ which produced the first magnetisation, and half a cycle of magnetisation will have been done.

Again the value of $-H$ is decreased to zero, and a residual $-B$ will be found which will be of the same magnitude as the residual $+ B$. Again is H reversed, becoming positive in direction, and so on to the completion of the cycle.

The results of such an experiment are shewn on a curve by Fig. 81. The dotted line OMS represents the magnetisation curve starting with an unmagnetised piece of iron.

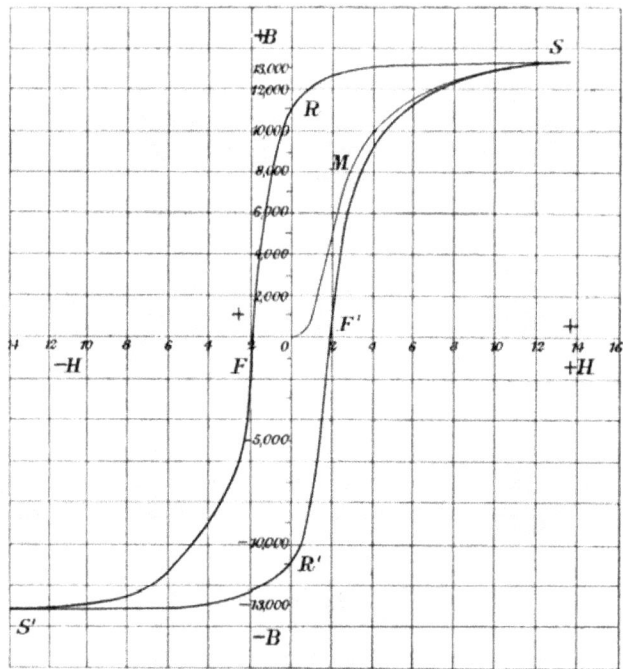

Fig. 81.

Saturation is reached when H is represented by $+ 13$. Then the part of the curve SR represents the magnetisation remaining when H is gradually reduced. It does not come back along the line of magnetisation; it lags behind, tending always to retain its last condition. The ordinate OR represents the residual degree of magnetisation when H has been reduced

14—2

again to zero. Then H is reversed and B is reduced and becomes zero when H is -2. Thus the abscissae OF' represents the *retentivity*. Then on further increasing H in the $-$ direction the curve FS' represents the corresponding effect. The point S' corresponds exactly to the point S. Then on again reducing H, OR' represents the residual in the other direction and is equal to OR. And again OF' corresponds to OF and the curve $F'S$ completes the cycle of magnetisation. It will be noted that the last quarter of the cycle (viz. $F'S$) does not coincide at all with the magnetisation curve OMS, except at the point S where saturation is obtained.

This curve is known as a *hysteresis* loop, for it shews clearly the tendency of the iron to lag behind throughout the cycle of magnetisation and to persist in its previous condition. And the tendency is called *hysteresis* (from the Greek : to lag behind).

Different samples of iron and steel exhibit this phenomenon of hysteresis in different degrees, and comparisons are best made by plotting out their hysteresis loops on the same scale under equal conditions.

Fig. 82 shews the cyclic curves for iron and steel under equal conditions. It is seen that, for equal values of H, B is greater for iron than for steel ; the residual of steel is greater in proportion (and is in fact in this case) than that of iron. It is also seen that the retentivity of the steel is greater than that of the iron, that the hysteresis is greater—the tendency to lag behind is greater.

It can be seen that if there were *no* hysteresis there would be no *loop* in the cyclic curve of magnetisation. The values of B on a decreasing H would correspond exactly to those on an increasing H, with the result that the cyclic curve would be a single line—a magnetisation curve extended on each side —not enclosing any area.

Now if this happened it will be seen that the work done in the process of magnetisation would be given out again in demagnetisation. The energy of a magnet must be represented by the amount of work necessary to set its molecules

in line. If they return entirely to their original positions
they would give back this energy in its entirety. But they
don't. Indeed, it is seen that to demagnetise the iron it is

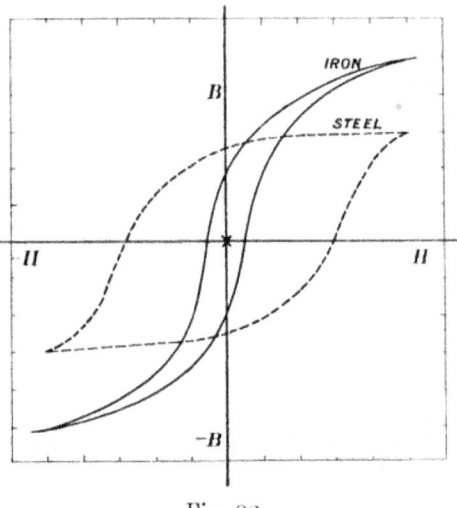

Fig. 82.

necessary to apply a reversed force—it is actually necessary
to do work in order to make the iron give up its energy.
And it can be seen that this work is in a manner wasteful—
and that it will depend upon the degree of hysteresis shewn
by the substance. The area of the space enclosed by a
hysteresis loop represents the *wasted* work done in the cycle
of magnetisation—by *wasted work* meaning the work done to
overcome retentivity or molecular friction or whatever cause
may be assigned to hysteresis. This may be reasoned out by
the reader. The wasted work is shewn as heat—witness the
heating of the iron core of an alternate current transformer.

The point of this discussion is that for practical purposes, if
iron be wanted for electromagnets, field-magnets, transformers,
or other electro-magnetic purposes the aim should be to have
it of high permeability and with small hysteresis. If it is

wanted for permanent magnets it should have large hysteresis and as high a permeability as possible. And the tests of comparison are to be found in cyclic magnetisation loops under equal conditions. The large-area loop shews marked hysteresis, the small-area loop is the curve of that material which will be chosen for use in alternate current work, or for armature cores, where reversals of magnetisation are inevitable. Hysteresis losses have always to be reckoned with : and like ordinary frictional losses, though they cannot be entirely obviated, should nevertheless to reduced to a minimum.

Variation of μ with Temperature. The permeability of iron and steel varies also with the temperature. Such variations are not uniform, but generally speaking it is found that with *small* magnetising forces the permeability is increased with an increase in temperature ; whilst with larger magnetising forces the permeability decreases with an increase in temperature. However, these statements only apply to temperatures below about 775° C., for it is found that the permeability of iron decreases suddenly in this region of temperature, and at 785° C. becomes unity. That is to say the permeability of the iron becomes the same as that of the air—and it loses its magnetic properties. The most powerful electromagnet will not attract a piece of iron at this temperature—or above it—which is called the *critical* temperature.

The following table shews the value of the permeability of a sample of iron under different magnetising forces H and at different temperatures.

Magnetising force H	μ at 21° C.	μ at 500° C.	μ at 775° C.	μ at 785° C.
0·5	2000	4000	8000	1
1	2000	5000	6000	1
2	4000	4000	3750	1
3	3700	3300	2400	1
4	3120	2560	2125	1
5	2600	2100	1700	1

It is seen from this table that the drop in μ between 775 and 785° C. is very sudden ; and it is also seen how the effect of temperature on the permeability depends also upon the magnetising force.

This remarkable loss of magnetic properties of iron at a high temperature is another support of the molecular theory of magnetism. At this critical temperature when the μ of the iron becomes unity other changes in the physical properties of iron take place. There is a change in the temperature-coefficient of resistance ; a change in the thermo-electric properties ; and a molecular change in the structure of iron which is evidenced by the fact that heat energy is absorbed at this temperature as the iron is being heated, over and above the heat energy necessary to raise the temperature, and is given out again on cooling. This phenomenon is known as *recalescence*. If a piece of thin steel wire (piano wire shews this most markedly) be heated to a bright red by passing an electric current through it and then be allowed to cool in a darkened room the glow will decrease until it is almost invisible and will suddenly brighten up again and disappear. This brightening is clearly the result of some internal evolution of heat—such as the giving out of heat by steam in condensing or by ice in melting. On heating the iron, it will be found that the temperature stands still for a little while at this critical point. Heat is absorbed, and that absorption is only due to some physical change at this temperature. Since the magnetic properties also change, the molecular theory is strongly supported by the phenomenon.

All substances conduct magnetic forces, but not to the same degree. Hence all substances have a *magnetic permeability*. But all substances are not capable of being magnetised, in the sense of retaining the properties of a magnet. Indeed there are only four substances known which are capable of magnetisation, viz. iron, steel, nickel and cobalt. These are frequently called *ferro-magnetic* substances, for they conduct magnetic forces so vastly better than the rest of matter. Iron has the highest permeability of any known substance, and it is closely followed by steel. Then nickel and cobalt—much lower in the scale yet so much higher than all other things, the permeability of which only differs slightly from that of air which is taken as unity.

Bismuth is the worst conductor of magnetic forces known and its permeability is 0·9998, very slightly less than that of air. The reader should be aware of the classification of bodies into *para-magnetic* and *dia-magnetic* groups. The first-named tend to set along a magnetic field and the latter across the field. These tendencies would follow from the facts that they conduct magnetic forces differently. The para-magnetic bodies have a greater permeability than air ; the dia-magnetic have a lower permeability. These imaginary lines of force will tend to crowd into good conductors which offer easy paths ; and will tend to avoid the bad conductors.

The following are some values of the permeability of certain substances under approximately equal conditions :

Substance	Permeability
Iron (Soft)	2500
Steel (Mild)	2000
Steel (Hard)	250
Cobalt	225
Nickel	200
Air	1
Bismuth	0·99982

But it must be remembered that these figures cannot be taken as "constants," nor do they represent relative permeabilities under equal conditions. They merely represent the permeabilities of particular samples under one set of approximately equal conditions.

The Magnetic Field about the Earth. The magnetic field about the earth is a factor which has to be taken into account in most magnetic measurements. It acts also as a sort of standard of comparison, because at any place it is uniform and constant over a considerable period of time. At any rate it is constant enough for most magnetic determinations.

It will be remembered that the magnetic field about the earth is somewhat similar to that which would be produced if a magnet were placed at the centre of the earth with its axis inclined to the earth's axis of rotation ; and that as a result the direction of the lines of force on the earth's surface change from a vertical to a horizontal as the equator is approached from a pole along one line of longitude. This, of course, is approximate, for the reader is aware that the earth's magnetic field is somewhat irregular.

Fig. 83 illustrates the hypothetical magnetic field about the earth assuming it to be produced by a magnet inside as

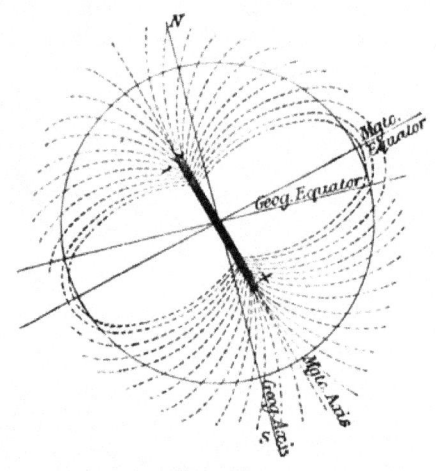

Fig. 83.

shewn. This will serve quite well to illustrate the main points involved by the effect of the earth's field upon magnetic measurements generally.

At the north magnetic pole of the earth the lines of force are in a vertical direction, acting downwards. As the circumference of the earth is traversed to the magnetic equator the direction of the lines become gradually less inclined to the horizon, until at the magnetic equator they are horizontal.

Continuing to the south the lines become more inclined until at the south magnetic pole they are again vertical but in an upward direction.

This is the first point then, that the direction of the earth's force in a given plane is different for every point on the circumference—that is for every point on one line of longitude. This is expressed by saying that the *angle of dip* varies continually on a line of longitude, the angle of dip being the angle between the direction of the earth's lines of force and the horizon at a place.

The second point concerns the strength of the earth's field at any point. This is not equal at all points—it is greater at the poles than at the equator, but the variation is *not* very great. Now by the strength of the earth's field at a place one means the number of lines per square centimetre at that place *at right angles to their direction.* This is more generally spoken of as the *total intensity* of the earth's field at a place.

Earth's Components. Now at a given place the total intensity acts in a given direction. In London, for example, that direction is at an angle of 67° 9′ with the horizon, in a vertical plane which is inclined at an angle of 16° 12′ W. of the geographical meridian at the present time.

A compass needle or a galvanometer needle is free to move only in a horizontal plane. It follows therefore that a compass needle cannot be acted upon by the whole of the earth's force at London. And it follows equally that at the N. magnetic pole of the earth it cannot be acted upon horizontally at all, since the earth's field is absolutely vertical. But in London, it *is* acted upon horizontally, but it is acted upon only by a *component force* of the total.

The total force can be resolved into two components—and for the greatest convenience they may be taken at right angles to each other, one acting horizontally and the other vertically. This is a mere application of mechanical principles.

On Fig. 84 a line *NS* is drawn whose length represents the magnitude of the earth's force and whose inclination to the horizontal AN is the angle of dip.   To resolve this force into horizontal and vertical components a line SB is drawn parallel to the horizontal AN and a line NB is drawn at

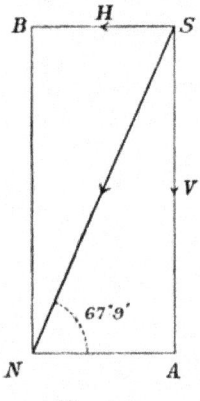

Fig. 84.

right angles to it.   These lines meet at B.   Then, according to the principle of the parallelogram of forces, the length of BN represents the *magnitude* of a vertical component, and the length SB the magnitude of the horizontal component of the total force along SN.

In other words the magnetic effect, both in magnitude and direction, at the point S due to two forces represented by SB and SA in magnitude and direction would be exactly the same as the effect produced by the force represented by the line SN.

It is seen from this that the magnitudes of the component forces H and V will always be less than that of the total force.   In London, H is less than V, and it is this H which acts upon a compass needle or upon a galvanometer needle and exercises a directive influence.

Now, assuming for a moment that the magnitude of the total intensity remains constant, it will be seen that as the angle of dip D becomes greater H will decrease and V will increase, though not in proportion.   And as angle D gets less H will increase and V will decrease.

At the poles the total force is vertical; there can be no horizontal *component* therefore. At the equator the total force is horizontal; there can be no vertical component therefore. Hence it follows on the assumption that the total force is constant, that as a compass needle is carried round the world the directive forces upon it will be *greatest at the equator* and *least at the poles*. (This is true as a matter of fact even allowing for the circumstance that the total force is greatest at the poles and least at the equator.)

Similarly it must follow that if an iron bar be carried vertically around the world, it will be magnetised by the vertical component and will therefore be magnetised to the greatest extent at the poles and not at all at the equator; and it will be taken through a cycle of magnetisation on a complete journey round a given line of longitude.

Referring back to Fig. 84, the magnitude of H will be given by

$$H = \frac{AN}{NS} \times NS.$$

But $\dfrac{AN}{NS}$ = the cosine of the angle of dip D,

and NS = the total intensity I.

$$\therefore \ H = \text{cosine } D \times I.$$

Similarly $V = \dfrac{AS}{NS} \times NS.$

and $\dfrac{AS}{NS}$ = the sine of angle D.

$$\therefore \ V = \sin D \times I.$$

At the present time the value of the total intensity I at London $= 0.46$; the angle of dip $= 67° \ 9'$; the horizontal component $H = 0.18$; and the vertical component $V = 0.42$.

Force on a compass needle in terms of H and tan a.   If the magnitude of the horizontal component, H,

of the earth's field be known at any place then the force
acting upon a compass needle at right angles to the direction
of H can be easily determined.

Fig. 85 represents a compass needle of pole strength m
and it is being deflected from its normal position NS by
means of some force acting parallel to EW.

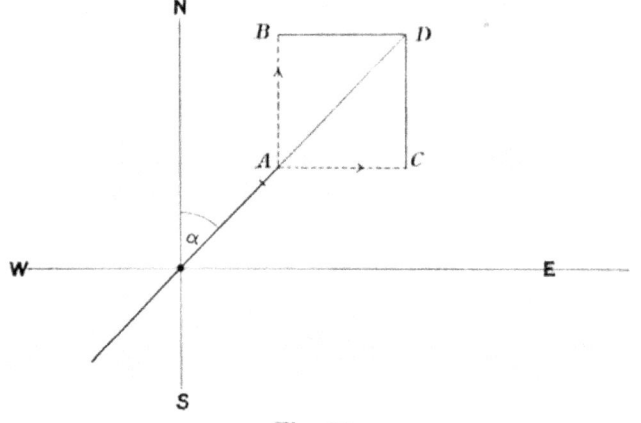

Fig. 85.

On the marked pole there are two forces acting. $m \times H$
represents the force due to the horizontal component of the
earth's field.   (H would be the force on unit pole, therefore
mH is the force on a pole m units.) mF represents the force
acting at right angles to NS at the place where the needle is.
The *resultant* of these forces will act along some intermediate
direction and the direction of the needle will be *along* it.
AD represents this direction.   It must follow therefore that
if DB be drawn from any point D parallel to AC, and if DC
be drawn parallel to AB, that

$$mF \, : \, mH \, :: \, AC \, : \, AB,$$
$$mF \, : \, mH \, :: \, BD \, : \, AB,$$
$$\therefore \; mF = mH \times \frac{BD}{AB},$$

$$\therefore \; F = H \times \frac{BD}{AB}.$$

But $\dfrac{BD}{AB}$ = tangent of the angle DAB, which is the angle of displacement on the needle a.

$$\therefore \; F = H \times \text{tangent } a,$$

where F is the strength of the magnetic field acting E and W *at the place where the needle is deflected.*

CHAPTER IX.

ELECTRO-MAGNETISM.

THE unit of current strength according to the absolute electro-magnetic definition is the strength of that current which flowing in a wire bent into a circle of 1 cm. radius produced at the centre a force of 2π dynes on a unit pole. Thus at the centre there will be a magnetic field of 2π lines of force per square centimetre.

If the current strength be C absolute units the strength of the field will be $2\pi C$ units.

The strength of the field at the centre of a flat coil varies inversely as the radius and directly as the number of con-volutions on the coil, the coil being of negligible thickness.

If there be n turns of wire on the coil, and its radius be r, then when a current of C absolute units passes through it the field at the centre will be

$$\frac{2\pi nC}{r} \text{ units.}$$

An *ampere* is $\frac{1}{10}$ of an absolute unit of current, therefore if C be in *amperes* the strength of the field at the centre will be

$$\frac{2\pi nC}{10r} \text{ units.}$$

In the case of a tangent galvanometer this field represents the deflecting couple, whilst the horizontal component of the earth's field represents the controlling couple.

When the needle comes to rest it has been shewn that deflecting couple = controlling couple × tangent of angle of deflexion, or

$$F = H \times \tan L.$$

Therefore in this case

$$\frac{2\pi n C}{10 r} = H \times \tan a,$$

$$\therefore C = \frac{10 H}{\frac{2\pi n}{r}} \times \tan a,$$

when C is in amperes.

The expression $\dfrac{2\pi n}{r}$ is a geometrical constant for a given coil depending solely upon the number of turns and the radius. This is sometimes called G,

$$\therefore C = \frac{10 H}{G} \times \tan a.$$

At a given place $\dfrac{H}{G}$ is constant, and this is known as the *constant* (sometimes electrolytic constant) of the tangent galvanometer, and is denoted by K,

$$\therefore C = 10 K \times \tan a.$$

That is to say

$$K \times \tan a = \text{current in absolute units,}$$

$$10 K \times \tan a = \text{current in amperes.}$$

The electrolytic method of determining K has been described on page 18. It is seen that if H be known at a place the value of K may be calculated by measuring the mean radius of the coil and the number of turns and reducing the expression $\dfrac{2\pi n}{r}$ to get G. Then $H/G = K$.

This should be done and compared with the result obtained by experiment.

Equivalent magnetic shell. When a thin plate of some magnetic substance is placed in a magnetic field so that the lines of force of that field act perpendicularly to it and consequently magnetise it perpendicularly to the plate it is called a *magnetic shell*.

The *strength* of that shell is measured by the product of the *thickness* of the shell (i.e. of the plate) and the *intensity of magnetisation*.

Now it can be shewn that the magnetic force exerted by a closed circuit in which a current of C *absolute* units is flowing is the same as that which would be exerted by a magnetic shell filling the area enclosed by the circuit and whose strength is numerically equal to C.

Strength of a magnetic shell = thickness of shell × intensity of magnetisation.

$$\therefore \text{Strength of shell} = l \times I,$$

since the thickness corresponds to the *length* of this thin magnet.

Now $\qquad I = \dfrac{\text{magnetic moment}}{\text{volume}} = \dfrac{2lm}{2la},$

where m = pole strength and a = area of cross-section.

$$\therefore I = \frac{m}{a},$$

$\therefore$ strength of each pole of the shell, $m = Ia$.

$\therefore$ the lines of force leaving one pole face of the shell and entering the other $= 4\pi m = 4\pi Ia$.

The Solenoid.   A solenoid is an elongated coil and may be regarded as being made up of a number of single thin coils placed parallel to one another on a common axis.

If a solenoid has N total turns and its length be L cms., and the radius of its cross-section be r cms., then

$$\frac{N}{L} = \text{number of turns per cm. of length} = n.$$

Therefore the thickness of each single coil (each single turn if there be but one layer of winding) $= \dfrac{1}{n}$ cm.

If a current of C *absolute* units be passed, then regarding each turn as an equivalent magnetic shell, the strength of that shell will be numerically equal to C and to the product of thickness and intensity of magnetisation.

$$\therefore\ C = \frac{1}{n} \times I,$$

$$\therefore\ I = Cn.$$

But there are N of these shells, all adjacent to one another, and the total effect of these will be the same as that of a cylindrical magnet of length L and of radius of cross-section r when magnetised to an intensity I.

$\therefore$ the total induction in this equivalent magnet $= 4\pi I a$, where a is the area of cross-section, and is equal to πr^2.

$$\text{Total induction} = 4\pi \cdot I \cdot \pi \cdot r^2.$$

But $\qquad\qquad\qquad I = Cn,$

$\therefore$ total number of lines in the cross-section of the solenoid

$$= 4\pi^2 r^2 Cn,$$

$\therefore$ the number of lines per unit area

$$= \frac{\text{total number of lines}}{\text{area of cross-section}},$$

$$= \frac{4\pi^2 r^2 Cn}{\pi r^2},$$

$$= 4\pi n C.$$

The total number of lines in the cross-section of a coil or solenoid or embraced by any electrical circuit is frequently called the *magnetic flux*.

The magnetic flux at any cross-section will thus be equal to the product of B (the lines per square centimetre) and the *area* of that cross-section.

Hence the strength of the field in the interior of a solenoid through which a current of C *absolute units* is flowing

$$= 4\pi n C \text{ lines of force per square centimetre.}$$

Therefore if a current of C *amperes* be passed through the solenoid, the strength of the field produced will be

$$\frac{4\pi nC}{10} \text{ lines per square centimetre.}$$

This latter is the expression in general practical use, C being in amperes.

The above expression for the strength of the field inside a solenoid is based on the assumption that air is the medium within the solenoid.

Now this magnetic field will be altered by the presence of any medium to an extent depending upon its permeability. If the interior be *filled* with iron then the strength of the field inside the iron will be

$$\frac{4\pi nC}{10} \times \mu \text{ lines per square centimetre,}$$

C being amperes and μ the permeability of the iron for the particular value of the magnetising force due to the current in the solenoid.

Magneto-motive Force. Inside the solenoid there is a uniform magnetic field except at the end portions. However, solenoids are generally made long compared with their diameter and the irregularity at the ends may be neglected.

Now a unit magnetic pole would be acted upon by a force of

$$\frac{4\pi nC}{10} \text{ dynes}$$

in a solenoid with air inside, when a current of C amperes was passing, n being the turns per centimetre. Now if the pole was urged through a length l centimetres inside the solenoid, the work done would be

$$\frac{4\pi nC}{10} \times l \text{ ergs.}$$

15—2

And if it was urged along the whole length L of·the solenoid, the work done would be

$$\frac{4\pi nC}{10} \times L \text{ ergs.}$$

It will be remembered that the difference of potential between two points is measured by the work necessary to urge a unit quantity of electricity from one point to the other. This difference of potential sets up a force—called an electromotive force.

It may be said then that the difference of magnetic potential between two points is measured by work done in urging a unit pole from one point to the other. This difference of magnetic potential sets up a magnetising force—called a magneto-motive force (M.M.F.). It may be compared to E.M.F. which can be regarded in the light of an "electrifying" force —a force tending to pass electricity from one point to another at a rate proportional to its magnitude.

M.M.F. is a magnetising force tending to magnetise substances to a degree of magnetisation proportional to its magnitude.

Reluctance. There is a law in magnetism resembling Ohm's law, that the total induction produced in any substance varies directly as the magneto-motive-force and inversely as the magnetic reluctance.

Reluctance corresponds to resistance. The magnetic resistance of a substance or its *Reluctance* varies directly as the length of the substance and inversely as the area of cross-section. It also varies inversely as the permeability of a substance. Thus, expressed in a uniform system of units,

$$\text{magnetic reluctance of a sub.} = \frac{\text{length of sub.}}{\text{area of cross-section} \times \text{permeability}},$$

or

$$R = \frac{l}{a \times \mu}.$$

The M.M.F. inside a solenoid of length L cms. and of

n turns per centimetre, when a current of C amperes passes round it, will be :

$$\text{M. M. F.} = \frac{4\pi CnL}{10},$$

but $n \times L$ will be the total turns on the solenoid, and may be denoted by S.

$$\therefore \ \text{M. M. F.} = \frac{4\pi}{10} \times SC.$$

Ampere-turns. The product SC is known as the *ampere-turns*.

It can be seen that a given magnetic effect can be produced by passing a small current through many turns or a large current through few turns. In practice, one desires a certain M.M.F. This is proportional to the *ampere-turns SC*. The value of this product is determined, and one can only settle which of the factors, S or C, shall be the greater by knowing the working conditions. For example, if the *ampere-turns* required for the *field magnets* of a *shunt* wound dynamo are known, then S is made as high as convenient and C as low as convenient. For a series wound machine, the whole current generated is to be passed round the field magnets. The number of turns S is regulated accordingly. These will be comparatively few.

Again, this helps one to determine the size of the wire which must be used in the winding. Obviously, small gauge wire when C is small and S is great, and *vice versâ*.

Measurement of μ : Magnetometer method. If a piece of iron of length l cms. and area of cross-section a sq. cms. be placed inside a solenoid, then the total induction (that is the total number of lines of induction cutting any cross-section),

$$N = \frac{\text{M. M. F.}}{R} = \frac{\dfrac{4\pi SC}{10}}{\dfrac{l}{a\mu}}.$$

$$\therefore \ N = \frac{4\pi \times S \times C \times a \times \mu}{10 \times l},$$

where μ is the permeability of the iron *for the particular magnetising force.*

This gives a means of determining the permeability by measuring N.

$$\mu = \frac{N \times l \times 10}{S \times C \times a \times 4\pi}.$$

Two methods will be described in this volume for the measurement of μ. This is the *magnetometer method.* The other, the *ballistic method,* will be described in Chapter X.

The apparatus used is identical with that described on page 199, used for the plotting of magnetisation and hysteresis curves. The apparatus is set up and compensation is duly made for the magnetic effect of the solenoid. The sample of iron is put into the solenoid, and a steady current of C amperes passed. The deflexion of the spot of light along the scale is noted, and should be *measured in centimetres.*

The various measurements to be made should be recorded as follows :

Strength of current in amperes in solenoid $= C$.

Total number of turns on solenoid . . $= S$.

Distance of lower end of iron from magnetometer needle . . . . $= p$ cms.

Perpendicular distance of magnetometer
needle from scale . . . . $= q$ cms.

Distance along scale through which spot
of light moved . . . . $= r$ cms.

Length of the sample of iron . . . . $= l$ cms.

Area of cross-section of the iron . . . $= a$ sq. cms.

Strength of horizontal component of earth's
field (or the field controlling the needle)
at the place where the magnetometer
needle is . . . . $= H$ units.

To illustrate this, the measurements of an actual experiment are given below, and the value of the permeability of the iron calculated.

Number of turns on solenoid . . . . = 800.
Current in solenoid . . . . = 0·6 amp.
Length of iron sample . . . . . = 50 cms.
Area of cross-section . . . . . . = ·0314 sq. cm.
Deflexion of spot of light . . . . = 40 cms.
Perpendicular distance of needle from scale = 100 cms.
Perpendicular distance of needle from
   bottom of iron wire . . . . . . = 16 cms.
Strength of the field *controlling* the needle = 0·18 unit.

Now $$N = \frac{\text{M. M. F.}}{\text{reluctance}}.$$

N is the total number of lines passing through the iron and therefore the number entering one pole and leaving the other. If m denotes the pole strength of the iron thus treated, then $4\pi m$ will be the number of lines of force of the pole and will therefore be the total induction N.

$$\therefore \ N = 4\pi m.$$

From this N may be determined. It has been shewn that when a needle is deflected through an angle, the deflecting force $F = H \times \tan \alpha$, where H is the magnitude of the controlling force.

Now in this case $H = 0·18$, and $\tan \alpha$ may be calculated, since the length of the deflexion and distance of the scale from the needle have been measured. But it must be noted that when a beam of light is turned through any angle by a mirror, the angle through which the mirror turns is only *one-half* the angle through which the light is turned.

Thus it would not be correct to state that the angle of deflexion was such that its tangent was $\frac{40}{100}$; but that the angle of deflexion is one-half of the angle whose tangent is $\frac{40}{100}$, viz. 0·4. The angle whose tangent is 0·4 is 21°. Therefore the actual angle of deflexion of the needle is $\frac{21}{2} = 10·5°$.

And the tangent of 10·5° = 0·1853.

Hence since $F = H \times \tan a,$

$\therefore\ F = 0\cdot18 \times 0\cdot1853$

$= 0\cdot033.$

Further, the force at the place where the needle is due to the magnet pole at the lower end of the iron under test,

$$= \frac{\text{strength of that pole}}{\text{square of its distance from the needle}}.$$

$$\therefore\ F = \frac{m}{16 \times 16}.$$

But $F = \cdot033.$

$$\therefore\ m = 256 \times \cdot033,$$

$$= 8\cdot45\ \text{units}.$$

Note that the effect of the upper end of the iron is being neglected. This, of course, must be taken into account in very accurate determinations : but with the iron 50 cms. long and placed as it is in this experiment, the effect of the upper pole will not be very great. The correction will tend to give a *higher* value for m.

Since $m = 8\cdot45\ \text{units},$

there are therefore $4\pi m$ lines, viz. $4\pi \times 8\cdot45$ lines of force proceeding from it.

$\therefore$ The total induction in the iron, $N = 4\pi \times 8\cdot45$ lines.

But $$N = \frac{\text{M. M. F.}}{R} = \frac{\dfrac{4\pi}{10} \times SC}{\dfrac{l}{a\mu}}.$$

$$\therefore\ 4\pi \times 8\cdot45 = \frac{4\pi \times S \times C \times a \times \mu}{10 \times l}.$$

$$\therefore\ \mu = \frac{4\pi \times 8\cdot45 \times 10 \times l}{4\pi \times S \times C \times a}.$$

$$\therefore\ \mu = \frac{4\pi \times 8\cdot45 \times 10 \times 50}{4\pi \times 800 \times 0\cdot6 \times \cdot0314},$$

$$\therefore\ \mu = 275.$$

This value for μ is, of course, not a constant for the particular sample of iron. It is only the value of the permeability of this sample when the value of B is 3380.

The reader should check this by determining the value of H in the solenoid, and having found it, he should get

$$\mu = \frac{B}{H} = 275.$$

CALCULATION OF AMPERE-TURNS REQUIRED TO PRODUCE DEFINITE MAGNETISING EFFECTS.

Complete Iron Ring. The case of a complete ring of iron forms a simple starting-point for definite calculations of this kind. A ring of iron, whose mean circumference is 75 cms., is to be magnetised so that $B = 12,000$ lines of induction per square centimetre. It is given that the permeability of this iron when B is 12,000 is 900, and that the area of cross-section of a slice cut out of the ring is a rectangle of 8 sq. cms., its sides being 4 cms. and 2 cms. What ampere-turns will be necessary to produce this result?

Total induction = no. of lines per sq. cm. × area of cross-section, *i.e.* $\qquad N = B \times a,$

$$\therefore\ N = 12,000 \times 8,$$

$$= 96,000 \text{ lines.}$$

Now $\qquad N = \dfrac{\text{M. M. F.}}{\text{reluctance}},$

$$\therefore\ 96,000 = \frac{\dfrac{4\pi}{10} SC}{\dfrac{l}{a\mu}}.$$

It may be stated here that $\dfrac{4\pi}{10} = 1\cdot257$, which for most practical purposes may be taken as $1\cdot25$. Again, since $1\cdot25$ cms. is almost exactly half-an-inch, the value of the

M.M.F. is frequently spoken of as the *ampere-turns per half-inch*.

$$\therefore\ 96,000 = \frac{1\cdot25 \times S \times C \times a \times \mu}{l},$$

$$\therefore\ S \times C = \frac{96,000 \times l}{1\cdot25 \times a \times \mu}$$

$$= \frac{96,000 \times 75}{1\cdot25 \times 8 \times 900}$$

$$= 800 \text{ ampere turns.}$$

Effect of an air gap. This example can be extended to illustrate the effect of an air gap in a magnetic circuit.

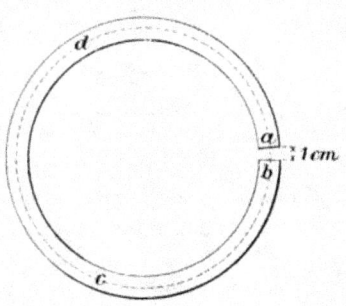

Fig. 86.

Fig. 86 illustrates the same iron ring, but a piece of iron 1 cm. long has been cut out as shewn. The mean length of the magnetic circuit is shewn by the dotted line *abcd*, and is, in this case, the mean between the inside and outside circumferences of the ring. This is not rigorously correct, but if the area of the section of the ring be small comparatively with the circumference it is sufficiently near.

It has just been shewn that with the complete iron ring 800 ampere-turns are needed to get a total induction of 96,000 lines through the iron. This effect may be produced by having 1600 turns with a current of 0·5 amp.; or 400 turns and a current of 2 amps.; or 20 turns and a current of 40 amps., and so on.

Now the magnetic reluctance of an air gap will be considerable compared with that of iron; for the permeability is only unity.

The total magnetic reluctance of the circuit, of which *abcd* represents the length, will be the sum of the reluctance of the iron path *bcd* and the air gap *ab*.

Let it be supposed that the same total induction, viz. 96,000 lines, is required. What must be the ampere turns necessary?

$$\text{Total induction} = \frac{\text{M. M. F.}}{\text{reluctance}},$$

$$N = \frac{\text{M. M. F.}}{\text{reluctance of iron} + \text{reluctance of air gap}};$$

$$\therefore \; 96,000 = \frac{\frac{4\pi}{10} SC}{\frac{l}{a\mu} + \frac{l'}{a'\mu'}},$$

where l', a', and μ' represent the length, area of cross-section, and permeability of the air gap respectively.

It will be seen that there must be some difficulty in defining the area of cross-section of an air gap. In this case the area of the opposite faces of the air gap is 8 sq. cms., but the lines passing from one to the other will *spread*, with the result that the nett equivalent area of cross-section of the gap becomes greater.

It has been shewn that in all cases the equivalent area of cross-section of an air gap is that of the opposite faces together with a fringe whose width is 0·8 × *the width of the air gap*.

Thus if the iron ring has a rectangular section of 4 × 2 cms. and the air gap is 1 cm., the equivalent cross-section of the air gap will be $(4 + 0·8) \times (2 + 0·8)$ sq. cms. If the air gap be 0·1 cm. the equivalent cross-section of the air gap will be 4·08 × 2·08 sq. cms.; and so on.

In this case therefore the reluctance of the air gap

$$\frac{l'}{a'\mu'} = \frac{1}{4\cdot8 \times 2\cdot8 \times 1} = \frac{1}{13\cdot44} ;$$

$$\therefore\ 96{,}000 = \frac{1\cdot25\ SC}{\dfrac{74}{8 \times 900} + \dfrac{1}{13\cdot44}} ,$$

from which $SC = 6501$ ampere-turns.

Thus to produce the same effect with the cut ring as with the complete ring 6501 ampere-turns are required instead of 800 ampere-turns. This means that magneto-motive force in the cut ring must be 8 times greater than that in the complete ring.

The reader should work out cases for himself, taking varying air gaps. He will find that even an air gap of 0·1 cm. causes an enormous increase in the total magnetic reluctance.

The moral should be obvious. In all iron magnetic circuits in which air gaps must exist, such as the field magnets and armature core of a motor or a dynamo, the air gaps must be kept as short as possible, for the shorter they are the smaller will be the M.M.F. required to maintain a given total induction through the circuit.

Magnetic Circuit. A magnetic circuit, such as those mentioned, may be made up of substances differing in length, area of cross-section and permeability. In such a case then the M.M.F. required to obtain a certain total induction N through the complete circuit will be the sum of the M.M.F.s necessary to get the same induction through each portion.

Or, $\text{M.M.F.} = NR_1 + NR_2 + NR_3 \ldots + NR_n,$

i.e. $\text{M.M.F.} = N\dfrac{l_1}{a_1\mu_1} + N\dfrac{l_2}{a_2\mu_2} + \ldots N\dfrac{l_n}{a_n\mu_n},$

where $R_1,\ R_2,\ R_3$ and R_n represent the reluctances of the several parts of the circuits. These might be, for example, the yoke, the magnet limbs, the pole pieces, the air gaps, and the armature core of a motor or a dynamo.

By application of this the ampere turns required for a given circuit and a given total induction may be calculated. In actual practice certain corrections must be made—allowances for *magnetic leakage*. It is found that although a certain N may be obtained in the yoke of an electromagnet, and although the M.M.F. applied is sufficient according to calculations, yet there is always a smaller induction than this through the armature core. Fig. 87 illustrates such a magnetic circuit. The difference between the induction in the yoke and in the armature core is due partly to *magnetic*

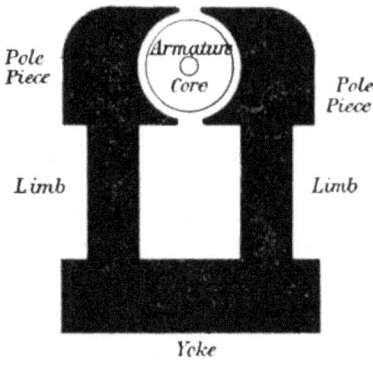

Fig. 87.

leakage, and partly to the magnetic conditions of the core itself, produced by the magnetic field of the current in the armature conductors. This will be dealt with later. But the question of magnetic leakage alone is one which will be readily understood. In a complete iron ring circuit there is no external evidence of its magnetic conditions—no stray lines of force. But the smallest air gap will cause leakage, i.e. stray lines of force other than those which pass between the faces of the gap. In the case of the circuit illustrated by

Fig. 87 investigations would shew that some lines pass between the limbs, some in the space between them and others on the outside. Such lines are of course wasted, and it is always desirable to reduce leakage to a minimum.

The limbs are generally bolted on to the yoke, and unless good magnetic contact be made by facing up the contact faces, there will be leakage as a result. The same applies to the pole pieces. Methods for the measurement of the leakage effect will be dealt with in Chapter X.

CHAPTER X.

ELECTRO-MAGNETIC INDUCTION.

THE production of an E.M.F. by the expenditure of mechanical energy was discovered by Joseph Henry and Michael Faraday independently. The results of their discoveries may be summed up by the statement that the magnetic conditions about an electrical circuit cannot be changed quickly. This is not obvious on the surface, but investigation of the phenomena shews that it is correct.

The production of the E.M.F. by the expenditure of mechanical energy depends upon this, for the energy is required to make any change in the magnetic conditions either by external or internal means.

Fundamental Phenomena. If a coil of wire be connected to a galvanometer and a pole of a bar magnet be thrust up to one face of the coil the galvanometer needle will be deflected. The deflexion will not be permanent, but after the first "kick" will oscillate uniformly about its zero position, shewing therefore that no current is passing through it.

When the magnet is removed again another deflexion will be produced, but this will be in the opposite direction to the first.

The E.M.F. induced by the motion of the magnet is called an induced E.M.F., and the currents are called induced currents.

Similar effects are obtained when a current is started and stopped in a neighbouring circuit. In Fig. 88 the circuit S

consists of a single turn and a galvanometer. The circuit P
is insulated from S entirely, and has a battery with a contact
key K in place of the galvanometer of circuit S. When a
current is started or *made* in P, an induced current will shew

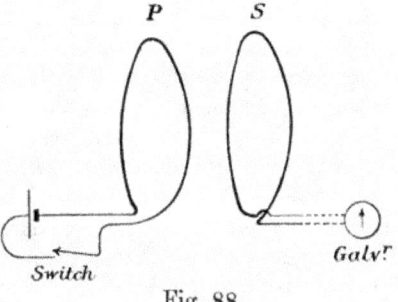

Fig. 88.

itself in S. When the current in P is steady there is no
effect in S. When the current in P is stopped or *broken*
another induced current shews itself in S, but in the opposite
direction to that obtained when the current in P was made.

Again, if the circuit S be moved nearer to or further from
the circuit P when a steady current is flowing in P, induced
effects will be obtained in S and their direction will depend
upon the direction in which S is being moved.

Again, if the strength of a current flowing in P be varied,
induced effects will be observed in S. These effects will have
one direction for an increased change in the current strength
and the opposite direction for a decreased change.

Again, if the circuit S be moved in the earth's field so
that by its motion it embraces a greater or lesser number of
lines of force, an induced effect will be obtained, the direction
of which will depend upon the nature of the alteration of its
magnetic flux.

These experiments all point to one thing, that the induced
E. M. F.s and resulting induced currents occur at those times
when a change is being made in the magnetic conditions of
the circuit. So long as the magnetic field about a circuit

remains unchanged there is no trace of induced effect. The steady current in the circuit P does not cause any induced effect; but when it is stopped, started, or altered in strength, effects are obtained. But those effects are not caused by variation of the current itself; they are caused by the variations of the magnetic effect of the current. At least one is led to this belief, since identical induced effects may be obtained by producing equal variations of the magnetic field about S by means of permanent magnets.

Let this be said then: that a variation of the magnetic flux of any electrical circuit sets up an E.M.F. of induction whose direction varies with the direction of the alteration. But that is merely a statement of observed fact, and though the word "causes" is used in that statement it must not be assumed that it is known *why* such variations produce such effects. One must be content for the present with the mere fact that these effects *are* obtained when magnetic variations are made.

Lenz's Law. The next point to determine is the relationship, if any, between the direction of the E.M.F. induced and the nature of the magnetic variation. This is done by experiment, and the result is that *the direction of an induced E.M.F. in a circuit is such that the magnetic effect of its resulting current will always tend to prevent the change in the magnetic flux of that circuit.*

This is known as Lenz's Law, and there is a certain inevitableness about it which must occur to the reader. Given the fact that induced currents are set up wherever a magnetic flux is altered it must follow that the magnetic effects of those currents will oppose and not help the change. For if the magnetic effect of an induced current helped the change which was the very cause of its production, then surely out of a small amount of energy there would grow an ever-increasing and unending quantity. For if the induced current helped the change that caused it, its own magnetic flux would cause a greater change, which would in turn cause a greater induced

E.M.F., and so on to infinity. This is against all knowledge and all known principles ; hence the inevitableness of Lenz's Law.

When the marked pole of a magnet is brought up to a coil some of its lines of force will thread the coil as shewn by Fig. 89 a. As the magnet is brought nearer so more lines will thread the coil. Now this change in the magnetic flux sets up an induced E.M.F. An induced current results and

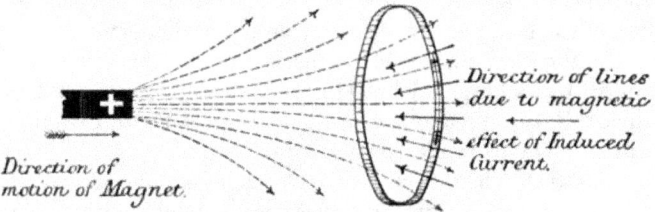

Direction of lines
due to magnetic
effect of Induced
Current.

Direction of
motion of Magnet.

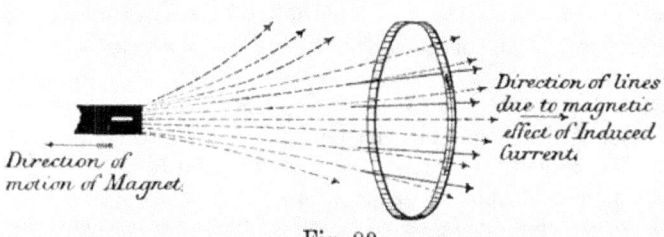

Direction of lines
due to magnetic
effect of Induced
Current.

Direction of
motion of Magnet.

Fig. 89.

its magnetic effect tends to oppose the change which is being made by the motion of the magnet. Hence the lines of force of the induced current will have a direction opposite to those of the magnet. The dotted lines indicate the lines of force of the magnet and the firm lines those of the induced current.

But when the magnet is moved in the other direction, tending to decrease the number of lines threading the coil, then again the induced effect tends to prevent the change, to prevent lines being removed. Then the induced E.M.F. is such that the magnetic effect of its resulting current sets up lines of force in the *same* direction as those of the magnet. Fig. 89 *b* shews this.

If the unmarked pole of the magnet be brought up and then removed the directions of the induced E.M.F.s will be the reverse of those given above.

If the magnet be brought up to the coil so that its length is parallel to the plane of the coil and its centre on the axes of the coil, then since each pole will tend to produce equal and opposite effects there will be no alteration in the magnetic flux of the coil, and consequently no induced effects.

In the case when the magnetic flux in the circuit *S* is altered by making and breaking a current in a neighbouring circuit *P*, the directions of the induced E.M.F.s and resulting currents may be similarly reasoned out.

On *make* in the circuit *P*—which may be called the *primary* circuit—an induced effect will be obtained in circuit *S*, the *secondary* circuit, whose magnetic effect will oppose the direction of the magnetic effect of the primary current. When the primary is broken then the induced current has such a direction that its magnetic effect has the *same* direction as the magnetic field of the primary current. That is to say it tends to prevent any change in its new magnetic condition. This must be thoroughly understood : that the magnetic field of the induced currents does not necessarily oppose the magnetic field which is being altered : but it always opposes *the alteration* : it tends to preserve the *existing* magnetic conditions *whatever they may be*.

The Source of the Energy of an induced Current.

Now it will naturally be asked, what is the source of the energy of these induced currents? It has already been pointed out that a bar magnet does not of its own account give out energy unless in the process it becomes demagnetised. When a marked pole is brought up to a coil face, lines of force are set up by the induced current which tend to oppose the motion of the magnet. Thus more work will be required to bring the magnet up to the coil than would have been required had the coil been on open circuit, and this additional energy is the energy of the induced current. When it is removed again, the lines of force of the resulting induced current tend to draw the magnet into the coil, and prevent its withdrawal. Thus more work has to be done in moving it away than would have been the case if no induced currents had been generated. This additional energy again represents the energy of the induced current. The energy is thus mechanical and *not* magnetic, although the mechanical energy has to be expended in order to overcome magnetic forces.

Having agreed to the above, another question will arise. Where does the energy spring from in the case of induced currents in a secondary circuit due to a make or break of a current in a primary? This does not admit of so simple and straightforward an answer, and one can only point out to the reader at this stage of his learning that when a current is started in a wire, a magnetic field is "built up" as it were about the wire. This field represents potential energy, and whilst the current is growing this energy is being accumulated. With a steady current, this potential energy stored in the surrounding space in the form of a magnetic field remains unaltered, but when the current is stopped, the magnetic field "collapses," and it is but natural that one should expect its potential energy to be given out.

Distinction between Mutual and Self-Induction.

When induction effects are produced in a circuit as a result of the variation in its magnetic flux by some *external cause,* the phenomenon is termed *mutual induction.* The word *mutual* refers to the reaction of the magnetic effect of the induced current upon the magnetic field of that external source which is being used in order to produce the alteration. It has been shewn how this reaction sets up an induced E.M.F. in a primary circuit, this E.M.F. being in addition to the active E.M.F. which is urging the primary current through its circuit, helping or opposing it according to the direction in which the magnetic flux was being altered, and therefore according to the direction of the flux of the induced current in the secondary. It is seen then that there will be a mutual alteration of the magnetic conditions of the two circuits ; that they will react upon one another mutually.

But when a current is passed through an isolated circuit, the magnetic field about that current will cause the magnetic conditions of the circuit to be altered. Here is a case when the magnetic flux of a circuit is altered without any external cause : it is altered by the current which is passed through it. Nevertheless, at the moment of starting the current in this circuit, an induced E.M.F. will be set up tending to prevent such alteration.

How can it tend to prevent it ? By opposing the active E.M.F. of the battery or dynamo which is being used to urge the current round the circuit. In short, an E.M.F. of induction is set up in this isolated circuit, tending to prevent any change in its magnetic flux and therefore, at the moment of switching on the current, acting in the opposite direction to the applied E.M.F. This induced E.M.F. is called an E.M.F. of *self-induction*, since it is caused by the variation of the magnetic flux of the circuit by the current which is started in it, and not by any external cause.

The effect of this self-induced E.M.F. on "making" a current in a circuit is to prevent the strength of the current from reaching its normal value (according to Ohm's law) instantly. The self-induced E.M.F. is a "*back*" E.M.F. of gradually decreasing magnitude, and it is only when this has decreased to zero that the current in the circuit will have reached its full strength for the given conditions. In other words, the current will take an appreciable time to "grow" to its full strength as a result of this self-induction.

When it has reached its Ohm's law value then it will be steady, and the circuit will now have new magnetic conditions. So long as they remain steady, there will be no further self-induction in the circuit.

But on switching off again, the magnetic flux due to the current will be withdrawn : an alteration in the magnetic

conditions will be made.   An induced E.M.F. will again be
set up tending to prevent the alteration.   In order to do
this, it is clear that it must tend to keep the current in
the circuit : to retain the previous magnetic flux.   Thus the
E.M.F. of self-induction at the moment of "breaking" the
current will be in the *same* direction as the applied E.M.F.
in the circuit.

Effects of Self-Induction.   These, then, are the effects
of self-induction ; a back E.M.F. is set up on make, and an
extra or helping E.M.F. is set up on break ; and there is
no effect whatever when a steady current is flowing.   If the
current varies, however, then there will be a back E.M.F. of
self-induction when the current increases, and an extra E.M.F.
of self-induction when it decreases.

The extra E.M.F. at the moment of "breaking" a circuit
may be shewn by connecting up some "highly inductive"
circuit (such as an electromagnet) to a battery of, say,
12 volts.   If a 100 volt lamp be also connected to the
terminals of the circuit, as a shunt, it will not light up
with the applied E.M.F. of 12 volts.   When, however, the
current is switched off, the lamp will be observed to glow
at the moment of breaking, and this may be made more
apparent by rapidly switching on and off.   At each break,
the extra E.M.F. of self-induction will be sufficient to urge
a current through the lamp great enough to produce incandescence.

The back E.M.F. of self-induction at make is not so readily
shewn by experiment.   The period during which the current
is "growing" is not a long one.   It is shorter, for example,
than the time required for an incandescent lamp to light up
fully after the switch has been put on.   It is shorter than
the time required for the indicator of any current measurer
to come to a position of rest.   Thus the phenomenon is difficult
to shew directly.   The following experiment serves the purpose however.

Two equal resistances, one highly inductive and one non-

inductive, are connected up to the same source of continuous
E. M. F.—that is to say in parallel with each other—and controlled by one switch. In series with *each* of these equal
resistances, an equal incandescent lamp is connected. When
the switch is put on, it will be seen that the lamp in the
non-inductive branch attains its maximum brightness some
time in advance of the other lamp ; although, of course,
ultimately both lamps will be equally bright. A large electro-
magnet will serve admirably as the inductive resistance,
and an equal resistance coil of platinoid wire will have a
sufficiently small inductance to be used as the " non-inductive " resistance.

Inductive and Non-Inductive Circuit. An inductive
circuit is any circuit in which the magnetic flux will be altered
by the passage of a current through it. Some circuits are

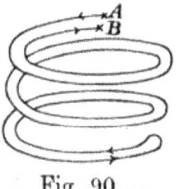

Fig. 90.

more inductive than others. A coil is much more inductive
than an equal straight wire circuit, although it should be
remembered that a straight wire circuit is not a possibility.
A coil of many turns is more highly inductive than one of
few turns, because a given current will produce in it a greater
magnetic flux. If iron be placed inside a coil, its *inductance*
is similarly thereby increased. The *inductance* of a circuit
then may be measured by the magnitude of the magnetic
flux in it when a unit current is flowing. The greater this
is then, the greater will be the alteration in it when a current
is started or stopped, and therefore the greater will be the
self-induction effects.

A coil may be wound so that it is *non-inductive*. This is

merely another way of saying that a coil may be so wound
that when a current is passed round it the resultant magnetic
effect will be zero. Such a coil is shewn by Fig. 90. A
current entering at *A* and returning to *B*, will produce no
resultant magnetic flux through the coil. Therefore there will
be no inductive effects, on making or breaking the current.
The resistance standards in resistance boxes are, or should be,
wound in this way, to be non-inductive.

Magnitude of Induced E.M.F. It was shewn by
Faraday that the magnitude of the E.M.F. *of induction* (both
self and mutual) *was directly proportional to the rate of change
of the magnetic flux.*

This expression, "rate of change of the magnetic flux,"
may be looked at in two ways. Firstly, the time required for
the change of a given flux; and, secondly, the magnitude of
the flux changed in a given time. A circuit may have a given
flux. This may be reduced to zero in a certain time; or
it may be reduced in half the time. The *average* E.M.F. will
be twice as great in the second case as in the first. Again, a
circuit may have a given flux reduced to zero in a certain
time, and then may have twice the flux reduced in the same
time. The average E.M.F. will again be greater in the second
case than in the first.

The idea involved may be proved experimentally quite
easily. A magnet may be brought slowly and then quickly
up to a coil connected to an ordinary galvanometer. This
will give the effects of a given flux change in varying times.
A certain current may be made and broken in a primary
circuit, and the magnitude of the induced effect on a neighbouring secondary noted. The primary current can then be
doubled or halved, or altered *ad lib.*, and the corresponding
effects noted. This will illustrate the effects of changing
different fluxes in equal times.

It was further shewn that the induced E.M.F. in any circuit
depends upon the number of convolutions. If a given change
in the magnetic flux of a circuit consisting of a single loop

causes a certain E. M. F. it may be proved by direct experiment that if the circuit consists of two convolutions and the same flux change be made the induced E. M. F. will be doubled. The induced E. M. F. then varies directly as the number of turns of wire embracing the magnetic flux which undergoes the change, and directly as the rate of change of that flux.

Neumann (1845) gave this law definite expression and shewed that

$$\text{Average E. M. F. induced} = \frac{(N_1 - N_2) \times n}{T},$$

the E. M. F. being in absolute electromagnetic units, N_1 and N_2 being the number of lines of force before and after the change of flux respectively, n being the number of convolutions of the circuit in which the change was made, and T being the time, in seconds, required to make the change.

The E. M. F. must be given as an average, since the actual magnitude of the E. M. F. induced at any moment depends upon the rate of change at that moment, and since the rate of change will necessarily be variable, due to induction reactions.

The expression may be reduced to volts by putting 10^8 in the denominator of the right-hand side of the equation, since 1 volt $= 10^8$ absolute units of E. M. F.

Thus

$$\text{Average induced E. M. F. in volts, } E = \frac{(N_1 - N_2) n}{10^8 T}.$$

The average induced current C will $\therefore = \dfrac{E}{R}$, where R is

the *total* resistance of the circuit in which the induced E. M. F. is acting.

$$\therefore \text{ average current induced, } C = \frac{(N_1 - N_2) n}{10^8 \times T \times R},$$

C being in amperes and R in ohms.

From this the quantity of electricity which passes round the circuit in the time T may easily be found.

$$Q \text{ coulombs} = C \text{ amps.} \times T \text{ secs.}$$

∴ quantity of electricity which passes round the circuit in the time T secs.

$$Q \text{ coulombs} = \frac{(N_1 - N_2)\, n}{10^8 \times T \times R} \times T.$$

Thus T cancels out, and

$$Q \text{ coulombs} = \frac{(N_1 - N_2)\, n}{10^8 R}.$$

This expression is of great value for flux measurements.

The quantity Q may easily be measured by means of a ballistic galvanometer, R and n are also readily determined, and thus $(N_1 - N_2)$ may be calculated, for it will be found by

$$(N_1 - N_2) = \frac{Q \times R \times 10^8}{n}.$$

If it be known, for example, that N_2 is to be zero, then N_1 can be determined.

Ballistic Galvanometer for Measurement of a Quantity of Electricity. This is the underlying principle of the ballistic method mentioned in Chapter IX. for the measurement of permeability and the like.

The flux in a closed magnetic circuit cannot be determined by magnetometric methods. And there are many other magnetic field measurements which may be done more readily and with greater accuracy by employing a ballistic galvanometer and measuring the quantity of electricity discharged round a circuit by the induced E.M.F. caused by an alteration of magnetic flux.

It will be necessary here to digress somewhat from the discussion of electromagnetic induction to explain how a quantity of electricity may be measured by means of a ballistic galvanometer. The ballistic galvanometer has already been described on page 22, and it is there pointed out that the first kick of the heavy needle is proportional to the quantity of electricity discharged.

It is firstly necessary then to standardise the galvanometer, that is, to determine what quantity of electricity discharged through it will cause the needle to kick so that the spot of light will be deflected through the division on the scale. This is done by discharging a known quantity of electricity through the galvanometer, noting the deflexion produced by the first kick and calculating the quantity necessary to give only one division deflexion, on the assumption that the deflexion is directly proportional to the quantity discharged.

Condensers. For this purpose a *condenser* of known *capacity* is charged so that its plates are at a known difference of potential. A *condenser* may be said to consist essentially of two conducting surfaces parallel to and insulated from one another. A leyden jar is a condenser: Aepinus' condenser is another example: two sheets of tinfoil separated by a piece of paraffined paper would make another. The reader remembers that in the case of the leyden jar a large quantity of electricity had to be given to the inner coating in order to get it up to the potential of the charging source. This held if the other coating of tin foil was earthed. But if the inner coating was taken out and charged by itself, isolated from other bodies, only a very small quantity was necessary to raise it to the desired potential. The quantity was measured, or approximated, by counting the number of sparks which could be passed to it. The coating, isolated, has a small *capacity*. The coating, when placed near to the earthed outer coating and separated from it by the glass of the jar, is said to have a large *capacity*.

The *capacity* of a conductor is a measure of the *quantity* of electricity necessary to give to it in order to raise it to a given potential. A body with a large electrostatic capacity requires a large quantity to raise it to a given potential; whilst a body with small capacity only requires a small quantity to produce the same potential.

Now the inner coating of a leyden jar is merely a conducting surface with a small capacity which really depends upon its area. But when it is placed in conjunction with its outer coating and the glass jar, the whole arrangement is called a *condenser*. The glass and the outer coating may be looked upon as a means of increasing the capacity of the inner coating.

A typical condenser for general use is made of sheets of tinfoil and paraffined paper. These are laid in alternate layers—paper, tinfoil, paper, and so on—up to any even number of tinfoil layers. Then all the odd-numbered layers of tinfoil are connected together at one corner and all the even-numbered layers are connected together at another corner. Thus the condenser is equivalent to two large areas of tinfoil separated from each other by a dielectric of paraffined paper. The *capacity* of the condenser will depend upon the total area of the two sets of tinfoil "plates," and will be greater for a larger area.

When a cell is connected to the condenser plates, equality of potential will be established between the positive terminal of the cell and the plate to which it is connected ; and again between the negative terminal of the cell and the other plate. But of course, since the dielectric has practically an infinite resistance, there can be no permanent current in the circuit. At the same time, however, in order that these condenser plates shall be charged to the same potentials as the positive and negative terminals of the cells respectively, a certain quantity of electricity must pass to them. When two bodies at different potentials are connected together electricity will flow from the body at a higher, to the body at a lower, potential until the potentials of the two are equal. The quantity of electricity which must pass in order to produce this result will depend upon the *capacity* of the body. It may be assumed that the actual difference of potential between the cell terminals remains constant during all the charging processes, for the quantity of electricity necessary to produce the desired potential difference between two condenser plates is always extremely small in practice. Thus it may be said that the quantity of electricity which must pass to the condenser in order to get a given potential difference between its plates is directly proportional to its *capacity*. Further, with a condenser of given capacity the quantity will be directly proportional to the difference of potential desired.

Unit of Capacity. Thus it follows that, in order to charge a condenser of capacity K so that its potential difference is V, the quantity of electricity Q will vary as $K \times V$.

The practical unit of capacity is called a *Farad*, and *is the capacity of the condenser which requires one coulomb of electricity in order to produce a difference of potential of one volt between its plates.*

Thus $$Q = K \times V,$$

when Q is in coulombs, K in farads, and V in volts.

After charging the condenser, it follows that if the plates be connected together the same quantity of electricity which was required to charge will be discharged through the connecting wire or galvanometer, or whatever the circuit may be completed with.

A means is thus provided of discharging a known quantity of electricity through a ballistic galvanometer and determining the "quantity per division" of the deflexion.

Calibration of a Ballistic. To standardise the galvanometer, a circuit is connected up as shewn in Fig. 91. A "morse" key, the terminals of which are represented by D, E, and F, is generally used as a simple means of charging the condenser and then discharging it through the galvanometer without having to make any alteration of the connexions.

The principle of the key is illustrated by the side diagram. A bar of copper or brass is hinged at the end F, and a spring S keeps this bar in contact with a brass piece E, which is connected to a corresponding terminal. When the bar is pressed down at the end T it will make contact with another brass piece D, which is also connected to a corresponding terminal. The bar is shewn at a half-way position. When the key is down the terminals D and F are connected together ; when it is up the terminals E and F are connected together.

In the diagram of connexions, K is the condenser, BG the ballistic galvanometer, and RK a reversing key.

On following out these connexions it is seen that when the key is down, the cell is connected to the condenser, and the

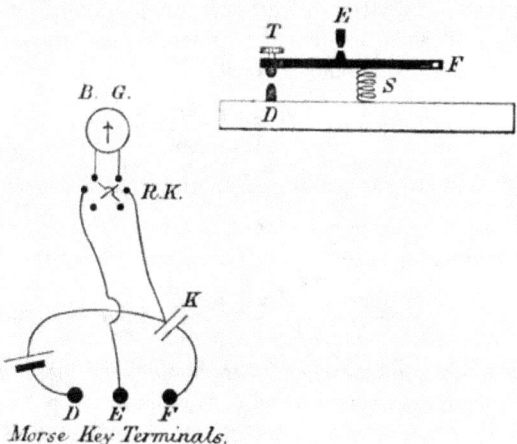

Morse Key Terminals.

Fig. 91.

galvanometer is cut out. This is charging the condenser. On letting the key up E and F are connected ; thus the condenser is connected to the galvanometer and the cell is cut out. This is discharging the condenser through the galvanometer. These connexions must be made strictly according to the diagram. The cell and K must not be transposed, and the connexion of one side of the galvanometer to the condenser must be made on the cell side as shewn.

The condenser is charged and then discharged through the galvanometer. The kick of the needle is read off and noted. This should be repeated some half-dozen times and consistent readings obtained. Care should be taken that the needle is *absolutely at rest* before discharging, otherwise, the needle being heavy, there will be large discrepancies in the results.

Let D be the mean of the deflexions obtained.

The quantity Q which produced this deflexion $= K \times V$, where K is in farads and V in volts.

Now a farad is too large a capacity for practice, so that condensers have a capacity of so many *micro-farads*, a micro-farad (usually written m.f.d.) being one-millionth of a farad.

Thus if K be expressed in micro-farads, then Q will be given in micro-coulombs. That is,

$$Q \text{ coulombs} = \frac{K \text{ micro-farads}}{1,000,000} \times V \text{ volts},$$

or Q micro-coulombs $= K$ micro-farads $\times V$ volts.

This quantity is calculated, K and V being known, and as it produces a deflexion of D divisions, therefore

$$\frac{K \text{ micro-farads} \times V \text{ volts}}{D} = \text{the micro-coulombs,}$$

which would produce *one* division deflexional kick. Thus the galvanometer is standardised. This quantity per division is generally given by the makers with the certificate of merit of the galvanometer (see page 24); but it should always be determined at the beginning of every experiment in which it is desired, because the conditions of the galvanometer may be different in the various cases. The controlling force may be different, the scale distance may be different, and so on. So that the makers' certificate is rather one which points out the capabilities of the instrument than a certificate of standardisation.

Magnetic measurements by Ballistic methods. A simple illustration of the measurement of a magnetic flux may now be taken in the determination of the number of lines of force which enter and leave a magnet. AB in Fig. 92 represents a long cylindrical bar magnet. S is a small "search coil" consisting of say 50 turns of wire. This

coil can be moved freely along the magnet, and it is connected up with long flexible wires to a Ballistic galvanometer
which has been standardised just previously. Now all the
lines of force which leave the pole B enter again at the pole
A, and therefore all pass through the cross-section of the
middle of the bar, that is through its magnetic equator.
Thus when the coil is at the middle position as shewn there
is a flux in it which is represented by the total number of
lines of force which leave the pole B. If the coil be quickly

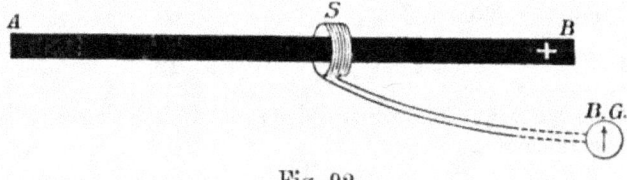

Fig. 92.

drawn off the magnet—*and drawn right out of the field of the
magnet*—to a place where the flux is practically zero, there
will be a change of flux represented by the flux at the centre
of the magnet. This change will set up an induced E.M.F.
which will cause a quantity of electricity to be sent round
the coil and galvanometer circuit. This quantity can be
measured by the galvanometer and from that the change of
flux in the coil determined.

In drawing off the coil it should be drawn well away
from the magnet pole and taken out to a place where the
flux in it is due only to the earth's magnetic field then the
value of N_2 is either known or may be in the majority of
cases taken as zero.

The coil should be put on again and a series of readings
taken, which should agree.

It may be supposed that a deflexion d was obtained.

Therefore the quantity of electricity discharged = d × the
quantity per division for the galvanometer in use = Q.

But $Q \text{ (coulombs)} = \dfrac{(N_1 - N_2) \times n}{10^8 \times R}.$

Now N_1 is the flux in the coil when it is at the centre of the magnet. N_2 is the flux in the coil when it is drawn right away, and in most cases may be taken as zero. n is the number of turns of wire on the coil S, and R is the *total* resistance of the coil and galvanometer circuit and will be the sum of the resistances of galvanometer, coil, and any other resistance which may be included.

Hence $$(N_1 - N_2) = \frac{Q \times R \times 10^8}{n},$$

and, if $N_2 = 0$,

$$\therefore \ N_1 = \frac{Q \times R \times 10^8}{n},$$

and since Q, R, and n are determined the value of N_1 may be calculated.

Worked Example. For example in a given case the quantity per division for the ballistic galvanometer was found to be 0.0055 micro-coulomb.

This is $0.000,000,005,5$ coulomb or 0.0055×10^{-6} coulomb.

The deflexion obtained by drawing off the search coil was 195 divisions.

Therefore the quantity of electricity discharged was

$$195 \times .0055 \text{ micro-coulombs}$$

$$= 1.0725 \text{ micro-coulombs} = 0.0000010725 \text{ coulomb},$$

$$= 0.00000107 \text{ to } \tfrac{1}{4}°/_{\circ}.$$

The total resistance of the coil and galvanometer circuit was 7000 ohms, and there were 50 turns of wire on the coil.

$$\therefore \ (N_1 - N_2) = \frac{.00000107 \times 7000 \times 10^8}{50}$$

$$= \frac{1.07 \times 10^{-6} \times 10^8 \times 7000}{50}$$

$$= \frac{1.07 \times 10^2 \times 7000}{50},$$

$$(N_1 - N_2) = 2 \times 1.07 \times 7000 = 14{,}980 \text{ lines of force.}$$

The value of N_2 may be neglected as the coil had a small area and was taken to a place where the flux density was only 0·18 lines of force.

Thus $N_1 = 14980$ lines of force.

This represents the number of lines leaving each pole. The result would have been the same in whichever direction the coil had been drawn off the magnet—although of course the direction of the deflexion would have been different.

From the results of this experiment the pole strength of the magnet may be determined. If m be the pole strength in terms of the unit pole then

$$4\pi m = \text{the number of its lines of force,}$$

$$\therefore \quad 4\pi m = 14980,$$

$$\therefore \quad m = \frac{14980}{4\pi} = \frac{14980}{4 \times 3\cdot142},$$

$$m = 1192 \text{ units.}$$

Again, if the distance between the poles be measured the moment of the magnet may be calculated, and is $2lm$, where $2l$ is the distance between its poles.

Magnetic Leakage Investigations. This same experiment can be utilised to determine the distribution of the magnetic field along the magnet. The ideal magnet is such that all its lines appear to diverge from a point at one end and converge to a point at the other. It would have no stray lines along its surface. But the real magnet does not even approach these ideals. The pole of a magnet, which is the region from which lines of force leave or to which they enter, instead of being a point or even an area at the extreme end, is extended along the magnet from one end to the magnetic equator. At the same time the bulk of the lines of force do enter and leave at the ends, but the flux density at the ends will never be as great as it is in the cross-section of the magnetic equator. Those lines which leave the magnet before they have reached the ends are known as *leakage lines*. Obviously the number of lines which leave a pole will be less than the number which cuts the cross-section at the equator by the number of these leakage lines.

If the search coil S be moved along the magnet from the equator in a series of jerks, each along a length of a centi-metre, for example, then the kick of the ballistic needle will be a measure of the change in the coil's flux. This change will be due to the leakage lines. In this way the leakage along each centimetre may be determined.

It is an instructive experiment, and if carefully carried out the sum of all the deflexions obtained by moving the coil one centimetre at a time and the deflexion obtained by taking the coil right off when the end has been reached will be equal to the deflexion obtained by drawing the coil off at once from the centre.

Further, if a piece of wood of the same section as the magnet be laid at one end, parallel to it, and the coil be

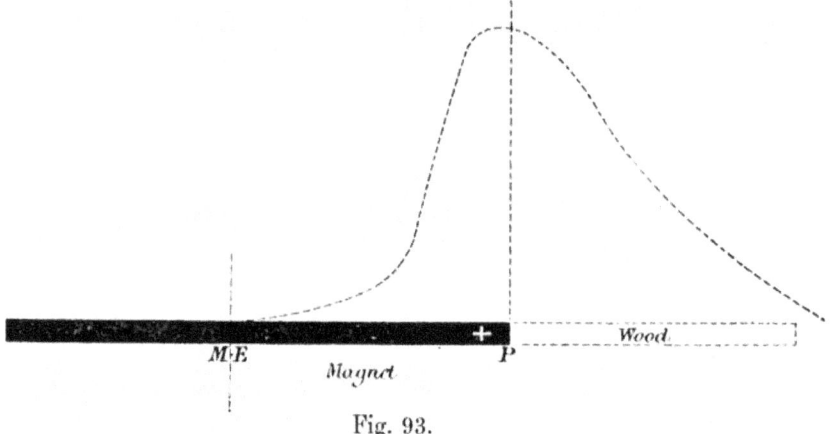

Fig. 93.

moved along it, still one centimetre at a time, a curve may be drawn up from the several deflexions taken which will illustrate the distribution of the magnetic forces along the length of the magnet.

The results obtained in an experiment may be shewn graphically as in Fig. 93. The ordinates represent leakage along each centimetre, and the abscissae are centimetres.

Ballistic method for determining permeability of an iron ring. A complete iron ring when magnetised in one direction around its circumference will not shew any external signs of its condition. If it is wound round with wire and a current be passed through, it will be magnetised to a degree depending upon the magnetising force and the permeability under the conditions of magnetisation. If the current be reversed in direction it will be magnetised to an equal degree in the opposite direction.

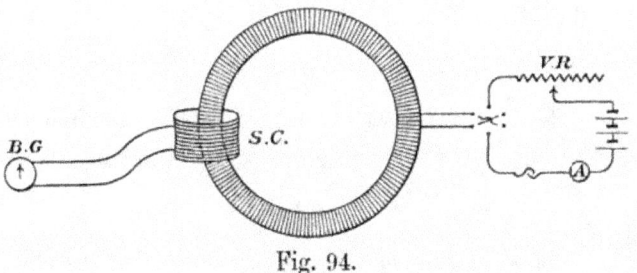

Fig. 94.

If a secondary coil—a search coil—be wound over the primary coil, or over a small part of it, then an induced E.M.F. will be set up in it whenever its flux is changed, that is whenever the magnetisation of the iron core is altered.

Fig. 94 represents an iron ring wound with a layer of insulated wire, the ends of which are connected through a reversing key, switch, variable resistance and ammeter to a battery of secondary cells or other constant source of E.M.F. A secondary coil, *SC*, is wound over a portion of this and connected up to a ballistic galvanometer.

If the ring is unmagnetised then when a current is switched on to the primary coil it will become magnetised and the flux conditions of the search coil will become changed from zero to

the flux in the iron ring.   Thus the effect could be measured
and the total induction in the iron calculated.   But this
could only be done *once* with a given iron ring, for on
breaking the primary current the ring would not become
demagnetised again.   In fact its condition would scarcely
alter at all.   But if the primary current be reversed, then
the magnetisation of the ring will be reversed so that its
total induction is the same but in the opposite direction.
This will change the flux in the coil from a given value in
one direction to an equal value in the other direction—that
is to say the change will be doubled.   Thus if N_1 represents
the total induction with the current of the primary circuit
in one direction, and N_2 the total induction for the same
reversed current, then if N_1 be positive in direction N_2
will $= N_1$ but be negative in direction.

$$\therefore \; N_1 = -N_2.$$

Therefore the change in magnetic flux in the iron will be

$$\{N_1 - (-N_2)\} = (N_1 + N_2) \text{ lines of force}$$

$$= 2N_1 \text{ lines.}$$

Hence in this, and in all cases in which the change of
flux is produced by the reversal of the current in the primary
coil, the change $(N_1 - N_2)$ is equivalent to $2N_1$ lines.

$$\therefore \; 2N_1 = \frac{Q \times R \times 10^8}{n},$$

and

$$N_1 = \frac{Q \times R \times 10^8}{2n},$$

where Q is quantity discharged by the effects of the change
of $2N_1$ lines of force in the secondary coil.

In this way then the total induction in the iron ring
produced by the magnetising force due to the current in the
primary circuit may be determined.

The permeability of the iron may then be calculated, by

determining the magneto-motive force, the mean length of iron ring and its cross-sectional area for

$$\text{Total induction} = \frac{\text{M. M. F.}}{\text{Reluctance}},$$

$$\therefore N_1 = \frac{\frac{4\pi SC}{10}}{\frac{l}{a\mu}},$$

$$\therefore \frac{Q \times R \times 10^8}{2n} = \frac{\frac{4\pi SC}{10}}{\frac{l}{a\mu}},$$

whence

$$\mu = \frac{Q \times R \times 10^8 \times l \times 10}{4\pi \times S \times C \times a \times 2 \times n} = \frac{Q \times R \times l \times 10^9}{8\pi \times a \times S \times C \times n},$$

where Q = the coulombs discharged in the secondary circuit, R the ohmic resistance of the circuit ; n the number of turns of the search coil, l the mean length of the iron ring in cms. ; a the area of cross-section of the ring in sq. cms. ; S the number of turns of the primary coil and C the current in the primary in amperes.

The variation of μ with the magnetising force may be determined by finding its value for different currents in the primary. Hysteresis loops can also be plotted out from results obtained starting with a large M.M.F. and determining the corresponding total induction. The M.M.F. is then reduced and the induction again determined and so on through the cycle. Of course the magnetisation must be reversed to get each reading.

Measurement of the effect of an air gap in a magnetic circuit. The same principles may be applied to measure the effect of an air gap in a magnetic circuit. This effect was discussed on page 234, and it may be determined by having a ring such as the above cut into two halves, the

ends being carefully faced up.   Each half should be wound
with insulated wire and the two windings should be joined
in series so that the magnetic effects of each are acting in the
same direction.   The two halves should be put together and
a secondary coil may be wound on either ; or for comparative
effects an equal coil on each, as shewn by Fig. 95.   With a

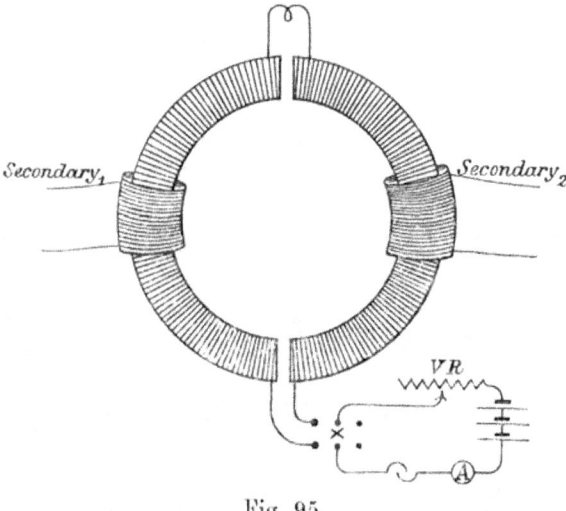

Fig. 95.

given current in the primary coils the total induction may be
determined when the halves are in contact.   Then the faces
can be separated with pieces of paper and the induction
determined.   Then pieces of cardboard can be put in between
the faces and so on.   In this way the effects of an air gap
on the total induction for a given M.M.F. are strongly brought
out.   The effect may be determined absolutely : or the
current in the primary may be altered in order to get the
same total induction for the various air gaps.   The deflexions
will be proportional to the total induction and the currents
in the primary will be proportional to the M.M.F.

The results of the experiments may be recorded as follows:

A. *Constant current in primary.*

Air gap	Kick produced on reversal of primary

B. *Constant deflexional kick—i.e. constant Total Induction.*

Air gap	Strength of current in primary

Investigation of Leakage in a magnetic circuit. The magnetic leakage of an electro-magnetic circuit may be investigated and measured by application of the foregoing principles. A simple case is illustrated by Fig. 96. A horse-shoe electromagnet provided with a keeper or armature, as it is often called, is connected to a battery through a reversing key. Three equal search coils are put on, one, S_1, on the *yoke*; one, S_2, on a *limb*; and the third, S_3, on the armature. These are connected through a 3-way key to a ballistic galvanometer.

When the primary current is on there will be a certain total magnetic induction. The whole of this will be at the yoke, which place corresponds to the magnetic equator of a bar magnet. Thus if S_1 be joined up to the galvanometer and the primary current reversed, a kick of the ballistic will be obtained, which will be proportional to the total induction at the yoke.

If there be no magnetic leakage then equal effects should be obtained when coils S_2 and S_3 are respectively joined to the ballistic and the primary current reversed. (S_2 and S_3 must have the same number of turns and the same resistance as S_1.)

If there is leakage, however, it will be found that the deflexional kicks differ ; and it may be predicted generally that the kick obtained for the limb coil will be less than that obtained at the yoke. The *difference* between the kicks will represent the *leakage*.

Again, at the armature coil the effect will be still less. The difference between the kicks for S_3 and S_1 represents the total leakage round the magnetic circuit. The difference

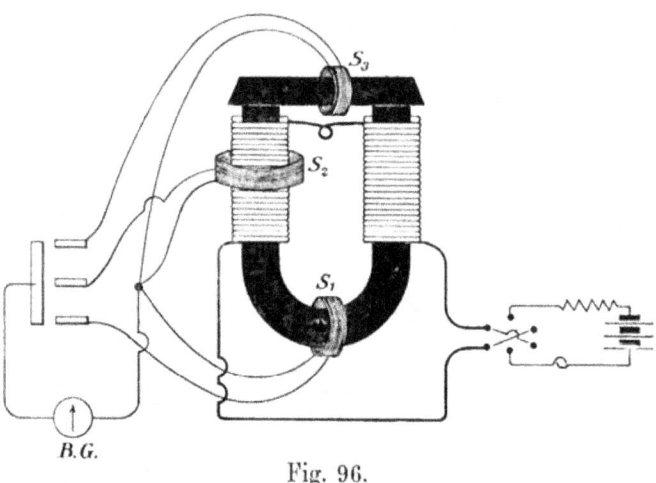

Fig. 96.

between the kicks for S_3 and S_2 represents the leakage between the limb at the place where the coil is and the armature centre. S_2 may be moved up or down the limb and the leakage between the yoke and each place noted.

The armature can then be separated from the limbs by different thicknesses of paper, and the effects of this on the total leakage may be determined. For each different gap, however, the effect at the yoke must be determined, and the effect at the armature must only be compared with the corresponding effect at the yoke. The experiment would be more rigid if the current in the primary were adjusted each time to get the same total induction at the yoke.

In a similar manner the leakage of the magnetic circuit of a dynamo or motor or transformer may be determined. It must only be remembered that to get any effect in the search coil used, the magnetising current *must* be reversed if the circuit is a closed iron one. When there are air gaps then the effects may be obtained by merely making or breaking the magnetising current. The magnetic circuit of a motor and dynamo will be discussed later.

Hopkinson's Bar and Yoke method. The ballistic method may be employed for bars as well as for rings of iron. But the leakage along an ordinary bar is liable to produce large

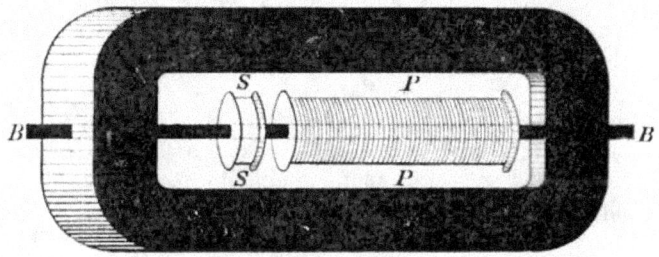

Fig. 97.

errors, with the result that special means have to be adopted to reduce this to a minimum. One method, due to Dr Hopkinson, consists in sinking the ends of the bar under test into holes in a large yoke of soft iron. The idea is illustrated by Fig. 97. The bar *BB* to be tested is put through the massive yoke as shewn, and on it are threaded two solenoids *S* and *P*. *P* is the primary through which the magnetising current will be passed. *S* is the secondary and will be connected to a ballistic galvanometer. The yoke forms a complete iron circuit, and being of high permeability iron, offering a very small magnetic resistance, it may be assumed that there is no leakage. Thus when a current is reversed in the primary coil *P* an effect will be obtained in the secondary *S* which will be a measure of the total induction.

Ruhmkorff Induction Coil. The Ruhmkorff induction coil may now be fairly considered as a practical application of electrical phenomena. Its function is the transformation of continuous E.M.F.s from a low to a high value. Its general construction is probably known to the reader. It consists of a primary solenoidal coil wound on an iron core of soft iron wires, and over this primary coil is wound a secondary coil the ends of which are connected to two highly insulated terminals placed as far apart as is practicable. The ends of the primary coil are connected to another pair of terminals through some form of current "break," by means of which a current in the coil will be rapidly made and broken.

For the moment the details may be disregarded. When a current is started in the primary coil there will be a change in its magnetic flux. This change will also be made in the flux conditions of the secondary coil. There will thus be set up an induced E.M.F. in the secondary coil. Again, when the current in the primary coil is broken, a flux change will be made which will set up another induced E.M.F. in the secondary, although it will be in opposite direction to the first. Thus at every make and break of the primary current there will be an E.M.F. of induction set up in the secondary, alternating in direction.

It may be considered that the change of magnetic flux in the secondary coil is approximately the same as the change in the primary. Neglecting other considerations then the relationship of the acting E.M.F. at the ends of the primary to the E.M.F. of induction at the terminals of the secondary will be simply proportional to their respective number of turns of wire.

In an induction coil the aim is to get a very high E.M.F. at the secondary terminals, high enough to send a spark across an air gap of length varying from half-an-inch to ten inches in general. Therefore the first point consists in having an enormous number of secondary layers and relatively few primary turns. Secondly, the insulation of the secondary windings must be very good, for between each layer there will be a considerable difference of potential. Thirdly, since a large change in the magnetic flux will tend to give higher E.M.F.s of induction, a M.M.F. of sufficient magnitude to raise the iron core to saturation point should be employed. To this end, since it must be advisable to keep down the number of turns on the primary, the wire must be of sufficient cross-sectional area to carry the necessary current without appreciable heating.

Fourthly, since the induced E.M.F. in the secondary depends not only upon the number of turns of wire or upon the actual change in its magnetic flux, but also upon the quickness of the change, it will be desirable to have some automatic "interrupter," which will make and break the primary current very rapidly.

Fifthly, since the amount of energy obtainable from the secondary circuit cannot possibly be greater than the amount put into the primary, it must follow that the strength of the current in the secondary circuit must be very small. For even under the ideal conditions, which are never realised,

E.M.F. of primary × primary current

= E.M.F. of secondary × secondary current,

at a given moment. Thus the wire used in the secondary winding need only be of small cross-sectional area.

These points must be given due effect in the construction of an induction coil.

There are various forms of "interrupters" or "breaks" for the purpose of making and breaking the primary current quickly and regularly. The spring trembling form is found on all small coils. It corresponds exactly to the make and break arrangement of an ordinary electric bell. Indeed, an electric bell, in series with the primary coil, would answer the purpose of interrupting quite well. Possibly, however, the bell would have to be wound to carry a larger current than usual.

There is the Wehnault chemical break, which consists of a vessel of sulphuric acid diluted with water (1 : 12), having a platinum wire as an anode arranged with its length at right angles to the plane of a kathode consisting of a large plate of lead. A high E. M. F. is needed when this break is used; but the current is interrupted very rapidly indeed. The interruptions are probably caused by the formation of gaseous films on the anode which disappear on break. These formations and disappearances take place very rapidly, so that the primary current is made and broken in rapid succession.

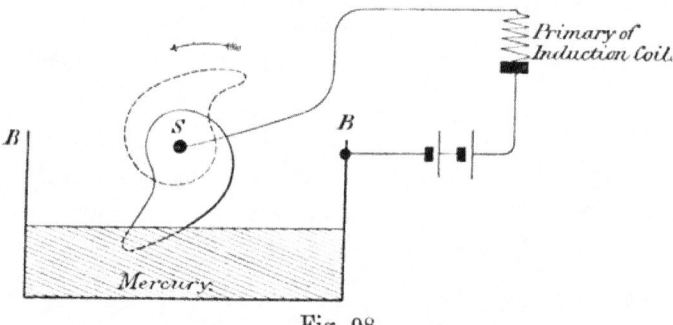

Fig. 98.

Another form of break is that known as the motor-mercury interrupter. This consists essentially of a cam-shaped piece of iron capable of rotation about a spindle S (Fig. 98). This spindle is rotated rapidly by means of a small motor separately driven. As the cam rotates it makes a series of contacts and breaks with the mercury which is contained in an iron box BB. The electrical circuit from the battery to the primary of the induction coil is completed through the driving spindle and the cam to the mercury and the iron box, and thence to the battery again. The speed of the motor can be altered, and a plunger arrangement enables the mercury in the box to be raised or lowered. A layer of water is poured over the surface of the mercury to prevent excessive sparking at the break and to prevent the consequent vaporisation of the mercury.

At every make in the primary there is an E. M. F. of induction in the secondary, and a reversed E.M.F. at every break. Thus there would apparently be an alternating E.M.F. at the secondary terminals if rapid and regular interruption of the primary be made. But this is reckoning without self-induction, and self-induction plays a very important part in the working of the coil.

Self-induction prevents a rapid "make": it causes a lagging of the current behind the E.M.F. So much so indeed that the E.M.F. at the secondary terminals for a make of the primary current is small compared with that produced at break.

At break the E.M.F. of self-induction is *with* the applied E.M.F., and there is a rush of current at the moment of break, and a consequently bigger flux at that moment which must die down to zero almost instantly. This change sets up a big E.M.F. in the secondary coil, but of opposite direction to the small E.M.F. of make.

It is this E.M.F. of break which is used. The comparatively small one of make is negligible. It will not cause a spark to jump across any air space worth considering. But the break E.M.F. will, and it is harnessed and used; the other is there, but it is not apparent. Thus it is that the induced E.M.F. at the secondary terminals appears to be unidirectional, and is unidirectional to all intents and purposes. For this reason it can be used to work X-ray tubes, which only work for a unidirectional discharge, and it can be used to charge leyden jars.

It is for this reason that the Ruhmkorff induction coil is often called a continuous current transformer, that is a means of transforming a low unidirectional E.M F. into a high unidirectional E.M.F. The reader will see that it is not an efficient machine, for only half the changes of magnetic flux in the primary are capable of being used—those changes produced at break.

The self-induction effects at break tend to make a large spark at the contact breaker. To take this up and utilise some of the energy a condenser of suitable capacity is con-

nected between the points where contact is broken.   This has
the effect of reducing the spark and making the break quicker.
It also retards the growth at make still more.

Eddy Currents.   Induction effects are by no means
confined to coils of wire.   They take place just as surely
within masses of metal.   If a pivoted needle be caused to
oscillate to and fro over a wooden table it will gradually come
to rest, because of the friction of the bearing and the resistance of the air.   If the same needle be caused to oscillate
through the same original angle over a plate of copper, it will
come to rest in a much shorter time.

If a penny be suspended between the poles of an electro-
magnet by means of a piece of thread ; and if the thread be
twisted up and released the penny will spin round rapidly.
But if the electro-magnet be made "live" it will pull up and
almost stop, as though it had suddenly been plunged into
some very viscous liquid like treacle.   On deadening the
magnet again, the penny will start off and soon get up its
former speed.

If a disc of copper be spun round in a horizontal plane
about a vertical axis, and a bar magnet be suspended above
it, the magnet will rotate in the same direction as the disc.
The magnet may be shielded from air currents by interposing
a sheet of glass between it and the disc, but the result will be
the same.

These three simple experiments are illustrative of the
production of what are called "eddy currents."   In all cases
the effects produced were caused by the magnetic effects of
induced currents.   The direction of an induced current is
always such that its magnet effect tends to prevent the flux
change producing it ; tends therefore to prevent any *motion*
which is the means of causing a change in the flux.

The pivoted needle swinging over the plate of copper
causes changes in the flux conditions of the parts of the
copper underneath it.   Local induced currents are set up in
the copper mass, each one tending to prevent a change ; each

one therefore tending to stop the swinging of the needle. These local currents in the mass of copper are called eddy currents, or sometimes Foucault currents; but their cause and effect is the same as those of any other induced current. They appear different because they are circulating in one mass of copper, a number of them possibly in one small area, hence the term " eddy " current.

In the case of the spinning penny, the flux conditions are being changed by the penny's rotation. Induced currents are set up in it tending to stop these repeated changes—tending therefore to stop its own rotation. Indeed additional work

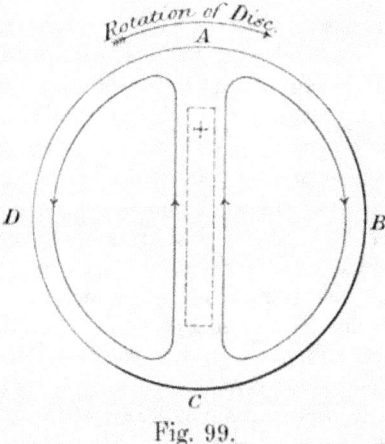

Fig. 99.

would have to be done to keep it rotating, and the eddy currents set up during that enforced rotation would heat the penny to a high temperature.

The case of the disc spinning with a magnet suspended above it illustrates the same thing. The magnet is originally at rest. The disc is turned and thus the flux conditions of its various parts become changed. Eddy currents are set up, and these tend to prevent the change; they tend to prevent the disc from being turned round; but since it is forcibly kept going, they will tend to drag the magnet round after the disc, always tending to get it to its own speed, for then, of course, there would be no flux changes in any part. This is known as Arago's disc experiment and a modification of it was devised by Faraday for the utilisation of some part of the electrical energy of the moving disc.

In Fig. 99 $ADCB$ represents the copper disc rotating as shewn. A bar is held over the disc in a fixed position shewn by the dotted lines. The direction of the resulting eddy currents in the disc are shewn. It will be seen that they are tending to prevent the change of flux in the various parts of the disc.

Along the diameter AC of the disc, parallel to the magnet, the E.M.F.s of both eddy currents are acting in the same direction. If a pair of copper brushes be connected to the terminals of a galvanometer, and one be held at A and the other at C so that they make a rubbing contact with the disc, the galvanometer will be deflected shewing that a difference of potential exists between A and C.

Moreover, for a given speed of rotation, the deflexion will be a permanent one. This is accounted for by the fact that although the disc is rotating and every part of it is being subjected to continual changes of flux, yet the direction of the eddy currents induced will not be altered at all with respect to the magnet though, of course, they are continually changing with respect to any one part of the disc.

This arrangement then may be called a *dynamo*—a means, that is to say, of maintaining an E.M.F. by the expenditure of mechanical energy. There is more energy required to rotate the disc under the stated conditions than would be required if the magnet were removed. But the arrangement is not a good dynamo; it is obviously an inefficient machine, since only a small fraction of the total electrical energy of the eddy currents can be used by "tapping" the disc at A and C. Nevertheless, it is a crude dynamo. As Faraday designed it the disc was rotated in a vertical plane about a horizontal axis, and it moved between the opposite poles of a horse-shoe magnet.   One contact was made at the axis of rotation and the other at the edge of the disc at that part which was passing between the magnet poles.

These eddy currents are set up in all masses of metal which are subjected in any way to alterations of their magnetic flux conditions.   They are not usually desirable.   They generally mean wasted energy, and that energy appears in the form of heat.   Means must therefore be taken to prevent them in all cases where they are not specially needed.

In the Ruhmkorff induction coil, for example, the iron core is not a solid mass of iron.   It is made up of soft iron wires, bundled up to form a cylindrical core.   Each wire should preferably be insulated from its neighbours, and it will be enough to dip the wires in shellac varnish to bring about this condition.   If the core were solid, it will be seen that when a current is made in the primary, induced eddy currents would be set up in the *core* in a plane at right angles to the core's length.   And again at break, there would be a similar set of eddy currents set up, but opposite in direction. These are useless—they are absorbing some of the energy which might be utilised—they are heating the core.   But if the core be made of the iron wires, then the cross-section will not be a continuous conducting surface of iron, with the result that eddy currents will not be set up.

That is a small illustration of the evil of eddy currents and the remedy for it.   In a dynamo or motor, the revolving parts are being subjected to rapid changes in their flux conditions.   These parts are made up largely of iron, the eddy currents which would be set up in them if they were continuous would always be tending to stop the rotations, and in addition would be heating the metal.

The following experiments illustrate the chief trouble of these currents and the means of preventing them.   Fig. 100 (*a*) represents a solid cylinder of copper suspended so that it can rotate about its vertical axis.   It is suspended by a thread which can be twisted up and which, when released again, will cause the cylinder to rotate due to its tendency to untwist.

Two pointers or indices PP are fastened to the cylinder to enable one to observe its rotation, and it is suspended between the poles of an electro-magnet which can be made "live" or "dead" as desired.

On releasing the twisted thread, having the magnet dead, the cylinder will soon attain a good speed. If the magnet be excited, however, the cylinder will be pulled up sharply, as though some strong break had been applied to it, and the torsional force of the thread will only be able to produce an

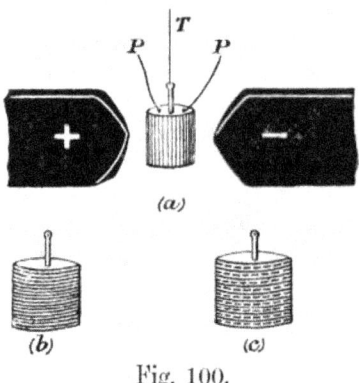

Fig. 100.

almost unnoticeable motion of rotation. On switching off the magnet current again, the cylinder will at once start off again with an increasing speed.

Next, a cylinder can be made up of circular copper discs, one upon the other, as illustrated by Fig. 100 (b), so that it has approximately the same dimensions as the solid cylinder. This may be called a laminated cylinder—laminated at right angles to its vertical axis. When this is treated to the same processes as the solid cylinder, it will be found that although it is visibly pulled up when the magnet is excited, it is not affected to the same extent. The torsional force of the thread can keep it moving, although at a greatly reduced speed.

Thirdly, a similar cylinder is built up with each copper

18—2

disc insulated from its neighbours by means of layers of thin paper. This is illustrated by Fig. 100 (c). The dotted lines represent the paper insulation, and the full lines the copper discs. This cylinder is laminated and insulated—again at right angles to its vertical axis. When this cylinder is suspended and spun between the magnet poles, it will be found that no appreciable alteration of speed is produced when the magnet is excited.

Lamination reduces the effects of the eddy currents, but lamination with insulation between the discs prevents them. This is the means taken in armature cores. Transformer cores are also laminated and insulated, but the insulation is usually secured by dipping the laminae in shellac varnish before building up the transformer.

It must be remembered, of course, that induced eddy currents will only be set up in a plane at right angles to the plane of the changing flux. Therefore lamination and insulation are only necessary at right angles to the plane of the induced currents. In the case of the cylinders of Fig. 100, no good would be done by building them up of laminae parallel to the vertical axis. This can be reasoned out by the reader.

Applications of Eddy Currents. These eddy currents have their uses however. In the hot-wire voltmeter described on page 38, it is pointed out that the instrument is rendered dead-beat by means of an aluminium disc fixed on the axis of rotation of the pointer. This disc moves between the opposite poles of a permanent magnet. As it does so, eddy currents are set up in it, and these always tend to stop the motion producing them.

In movable coil instruments, the coil is wound on a former of aluminium. This is continuous, and as the coil rotates induced currents are set up in the former, tending always to stop the motion. This is also the device for dead-beatness.

In the needle type of ballistic galvanometer, the needle

is suspended in a hole in a sphere of copper. Eddy currents set up by the motion of the needle tend to stop that motion. This copper sphere must be removed for some experiments.

In other needle galvanometers and in magnetometers, the needle often swings in a copper box, with the same object in view.

In the case of a suspended coil galvanometer, it will be found that the moving coil can be pulled up by short circuiting the terminals of the instrument. This may be done with a simple tapping key, and it will be found a very useful means of pulling up a swinging coil needle. This would apply chiefly to the bipolar suspended coil ballistic, which is not dead-beat, and in the using of which it is so essential that the needle shall be brought to rest before taking another reading. The short-circuited coil moving in the field of its magnet will have induced currents set up in it, which will tend to stop the motion producing them. In some cases it will be found that a direct short circuit pulls the coil up too much—hangs it up at a place and keeps it there. In such a case, if a resistance be put in series with tapping-key and the galvanometer more gradual, satisfactory results will be obtained.

Because of this hanging-up tendency of suspended coil galvanometers, they are not suitable for induction experiments —such as the ballistic method for measuring permeability— since the galvanometer circuit is a closed one with no permanent source of E.M.F. The kick obtained would be less than it should be, because of the hanging-up propensities of the moving coil.

CHAPTER XI.

DYNAMO PRINCIPLES.

A WIRE conveying a current tends always to rotate about a magnet pole. This is the underlying principle of the electric motor. A wire rotated about a magnet pole has an E.M.F. induced in it. This is the underlying principle of the dynamo.

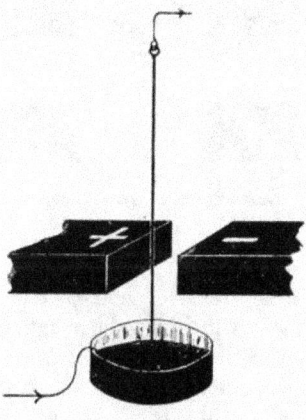

Fig. 101.

If a wire be suspended in the manner illustrated by Fig. 101, so that its lower end dips under mercury, a current of electricity may be passed up or down, and its lower end will be capable of movement within the confines of the vessel containing the mercury. Two opposite magnet poles may then be arranged so that a portion of the length of the wire lies between them, its length being at right angles to the magnetic field in the interpolar space.

A current may then be passed *up* the wire. A movement will be produced if the current and the magnetic field be strong enough. The wire will be observed to move *across* the field of the magnet poles, at right angles to their lines of force. It will come to rest at one extreme side of the mercury vessel. When the current is switched off again, it will swing back to its original central position.

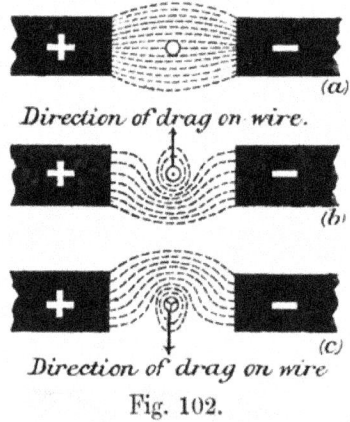

Fig. 102.

If the current be reversed and switched on, the wire will again move *across* the field of the magnet poles, but in the opposite direction to the first movement.

The resultant magnetic field of the magnet poles and the current in the wire will assist in the explanation of these effects. Fig. 102 consists of three diagrammatic plans of the section across the magnetic field between the two poles. The magnetic field due to the poles when no current is passing through the wire is shewn by (*a*). The resultant field when a current is passed *up* the wire is shewn by *b*. The dot in the centre of the section of the wire represents a current coming up. It is supposed to be the pointed end of an arrow, and is always used in this way to represent the direction of a current in a section of a wire. The opposite direction is represented by ⊗, and this is the feathered end of an arrow going away from the observer. The direction of the lines of force due to the current is shewn by the lines immediately about the wire. The distortion of the field between the poles is easy to predict, for the lines due to the current are in the same direction as the magnet's lines on one side, and in the opposite direction on the other side—bottom and top respectively in the figure (b). On the top side, the field will be weakened; and on the bottom side, it will be strengthened. Lines which would have gone straight across normally, now bend round the wire on that side of it where its lines of force have the same direction as theirs.

There is pressure across a magnetic field—lines of force of the same direction appear to repel one another—and the result is that the wire is acted upon by magnetic forces tending to push it across the field—upwards according to the diagram.

The figure (c) illustrates the case when the current is sent down the wire, and it is seen that the direction of motion of the wire across the field is reversed. If the direction of the magnet field between the poles had also been reversed, it can be reasoned out that the direction of motion of the wire would have been the same as in case b.

It can be seen that there will be no tendency for the wire to move in any way *along* the field, for the effects of lines curving around each side will be equal and opposite in direction.

The resulting force tending to drag the wire across the field is directly proportional to the strength of the field between the magnet poles; to the strength of the current in the wire; and to the length of the wire which is in the field.

The same apparatus may be used to shew the converse experiment. A galvanometer or other current indicator may

be substituted for the source of E.M.F. by which a current
is sent up or down the wire. If the wire be pulled across
the magnetic field in one direction a current will be indicated
by the galvanometer. If it be pulled across in the opposite
direction a reversed current will be indicated by the galvanometer. If it be moved *along* the field, parallel to the lines
of force between the magnet poles, there will be no indication
of any induced current ; only when the wire *cuts* the lines of
force is there any induction effect, for it is only then that
any change is being made in its magnetic conditions.

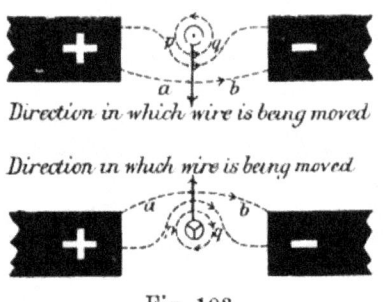

Direction in which wire is being moved

Direction in which wire is being moved

Fig. 103.

When the wire is moved in one direction across the field
the induced current set up will be in such a direction that it
tends to prevent the change : tends therefore to stop the
motion of the wire.

Fig. 103 is a diagrammatic sectional plan as before. The
direction of the field of the magnet poles is shewn by the
arrow *ab*. The direction in which the wire is being moved
is shewn by an arrow. Now an induced current will be set
up in the circuit of which the wire forms a part, and its
direction will be such that its magnetic effect tends to stop
the motion. Therefore on the side *pq* of the wire the lines
of force about the wire will be in the same direction as the
lines of the field, since there is repulsion between all lines
having the same direction. Hence the direction of the
induced current must be as shewn, namely *up* in Figure (*a*)

and *down* in Figure (*b*), and these facts may be verified
experimentally by determining the direction of deflexion of
the galvanometer needle for a known direction of current
through it.

It is clear from these diagrams then whilst the induced
current is flowing, the wire is acted upon by magnetic forces
tending to urge it in the opposite direction to that in which
it is being moved by the experimeter. This is made clearer
by comparison with the previous diagrams. Now the tendency
of the wire to be dragged in the opposite direction to that
in which it is being moved depends upon the strength of the
current, the strength of the field and the length of wire in
the field.

The induced E. M. F. set up in the wire, on the other hand,
depends upon and is directly proportional to, the strength
of the field, the length of wire in the field and the *rate at
which it cuts the lines of force.*

With a given resistance therefore the strength of the
induced current varies also with these things, since it must
vary with the induced E. M. F.

Thus it follows that as the induced E. M. F. is increased,
say by moving the wire quicker, the induced current must
increase and as a result the forces tending to pull the wire
in the opposite direction must also increase. *The work which
must be done in order to overcome the forces tending to urge
the wires in the opposite direction supplies the energy of the
induced current.*

Now it can be seen that as the conditions are so altered
that the current induced becomes greater, so more and more
work will be needed to overcome the resulting opposite forces.
It is because of this that a 100 horse-power engine, say, is
needed to turn the armature of dynamo generating a current,
whilst the same armature can easily be turned by hand when
it is on an open circuit.

Fleming's hand rule. A useful hand rule, due to
Dr Fleming, enables one to determine quickly the direction

of an induced current in a wire when it is dragged in one direction across a magnetic field. It can be used also, of course, to determine in which direction it must be moved to give a particular direction of current and so on.

The thumb, forefinger and centre finger of the *right* hand are extended in three directions, at right angles to each other —like the three edges which meet at the corner of a cube. The *thumb* represents the direction of *motion* of the wire; the *fore*-finger represents the direction of the lines of *force* of the magnetic field across which it is being moved; and the *centre* finger indicates the direction of the resulting induced current. If any two directions be known and the right hand so extended be held so that the respective fingers point in those directions, the direction of the remaining finger gives the direction of the third.

The reader can practice this on Figs. 102 and 103.

The rule may be used for motor directions, but the *left* hand must be used then, and the centre finger will denote the direction of the current which is passed through the wire. This may be practised on Figs. 102 (*a*), (*b*) and (*c*).

If a wire could be forcibly moved across a magnetic field so that it always travelled in one direction then a permanent and steady E.M.F. would be maintained, provided that the wire moved at a constant speed, across a uniform field, and with a fixed length in the field all the time. But it would be impossible to realise such conditions. The nearest approach to them lies in the Arago disc experiment and in that there is no wire but a mass of copper in which the greater portion of the energy generated is wasted away.

However, it is a simple matter to move a wire to and fro across a limited space in which there is a magnetic field. In such a case there would be a series of E.M.F.s generated, alternating in direction in the same way that the direction of motion of the wire is alternating. There would thus be a series of currents, in the circuit in which the wire formed a part, alternating in direction.

Not only would the E.M.F. be alternating; it would also

be varying, because in order to move a wire to and fro. it would be impossible to keep it going at a constant and uniform speed. It would have to be stopped at each extremity of the field in order to reverse its direction ; and since the E. M. F. varies as the speed of the wire across the field there would therefore be a variable and an alternating E. M. F. set up in this wire.

The most convenient method of moving a wire to and fro across a magnetic field is to fasten it along the side of a cylinder, parallel to its axis, and to arrange the cylinder so that it can be rotated about its axis. The cylinder can then

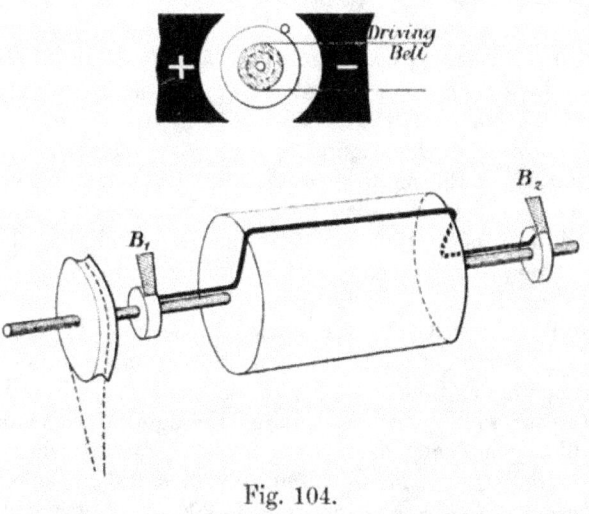

Fig. 104.

be put in the magnetic field so that its length is at right angles to the direction of the lines of force of the field. Now on turning it round continuously in one direction, the wire fastened on to its side is pulled across the field, first down and then up. The cylinder can be rotated at any desired speed and connexions to the wire can be made by having two metal rings on the axis of rotation—insulated from it—

one at each end. The ends of the wire can be soldered on to these rings, and two *brushes* of, say, copper gauze can make a rubbing contact on the rings. Fig. 104 illustrates the idea. B_1 and B_2 are the brushes making a rubbing contact on the insulated copper rings to which the ends of the wire, shewn by the thick black line, are soldered. This can be rotated between two curved pole pieces—as shewn by the end section at the top.

As this cylinder is rotated the wire moves to and fro across the field. It is not moving strictly at right angles to the field except at two points, half-way down the field and half-way up again. At the top and bottom of its motion it moves parallel to the field. It can be seen therefore that with a constant speed of rotation *the rate at which the wire cuts the lines of force is variable.* It is clearly zero at the top and bottom when the wire is moving parallel to the field, and it is clearly a maximum, for a given constant speed of rotation, at the midway positions in the field; and between these positions it varies, increasing from zero to maximum, and decreasing to zero; then increasing in the opposite direction to a maximum and decreasing again to zero. Such are the changes of the rate of cutting during one revolution.

But the induced E.M.F. varies as the rate of cutting, other things being constant; therefore during one rotation of the wire, starting from the top position, the E.M.F. will gradually increase whilst the wire rotates through an angle of 90°, will decrease again as it moves from 90° to 180° when it will be zero. On moving from 180° to 270° it will increase *in the opposite direction*, and from 270° to 360° it will decrease again to zero.

Variation of rate of cutting with a constant speed of rotation. It is necessary to determine how the rate of cutting varies with a constant speed of rotation in order to know the variation of the E.M.F. during one revolution.

The circle $AMNR$ in Fig. 105, represents the path of the wire during one rotation, and the direction of the magnetic

field is parallel to the diameter *RM*.   *AN* is at right angles
to this, therefore as the wire rotates it will be moving parallel
to the lines of force at *A* and *N*; and at right angles to them
at *M* and *R*.

Let the wire be moving round at a constant speed and
at a given moment let it be at the position *B*, having moved
through an angle *ACB* from the position *A*.

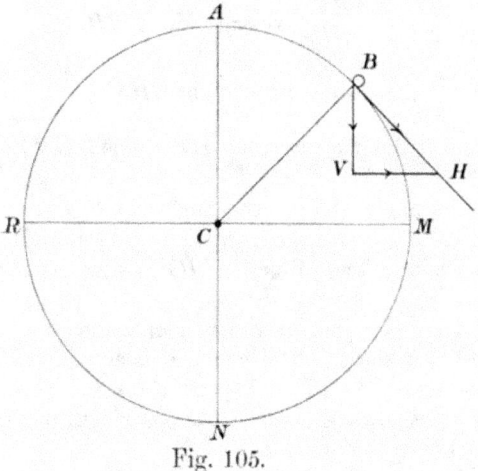

Fig. 105.

Now at this moment the actual direction of the wire is
a tangent to the circle at the point *B*.   Join *BC* and at
B draw *BH* at right angles to *BC*.   The direction along *BH*
is the direction of the wire's motion *at this moment*.

Now a length *BH* can be marked off to represent the
speed of the wire along this direction.   Any length will do
for the object in view.   Now this direction is *not* at right
angles to the field, but it will be simple to determine the
equivalent velocity of the wire across the field, at this moment,
by applying the principle of the triangle of velocities.

From *B*, draw *BV* at right angles to the lines of force
of the field and therefore to *MR*.   From *H* draw *HV* parallel
to the lines of force of the field.   These lines meet at *V*.

The velocity represented by BH in magnitude and direction has been *resolved* into two—one along BV and the other along VH, and the lengths of these lines so drawn represent the *magnitude* of these resolved velocities on the same scale that BH represents the velocity along BH.

Now the magnitude of the resolved velocity along BV, and at right angles to the lines of force,

$$= \frac{BV}{BH} \times \text{velocity along } BH.$$

But $\qquad \dfrac{BV}{BH} = \text{sine of the angle } BHV.$

∴ velocity along direction $BV = \text{sine} \angle BHV \times \text{velocity}$ along BH.

Now the angle $BHV =$ the angle ACB, since BH is at right angles to BC and VH is at right angles to AC.

∴ velocity along direction $BV = \text{sine } ACB \times \text{velocity}$ along BH.

That is to say, just at the moment when the wire is at the position B the speed at which it is actually *cutting* lines is only a fraction of its actual speed of rotation (linear speed), and this fraction is the sine of the angle through which it has turned from the position A.

Now for any other position of the wire the result will be the same, that the speed of moving across the lines of force is given by the product of the sine of the angle through which it has rotated from A, and the actual speed of rotation.

When the wire is at M, it is actually moving across the lines at right angles. It has moved through 90°. The sine of 90° is 1. Therefore the speed of cutting is the speed of rotation for this position. At the position A, there can be no resolved velocity across the lines. The angle is 0°, and sine of 0° is 0. Therefore the speed of cutting is zero.

But for a constant speed of rotation it is clear that whilst the actual speed of cutting $= \sin a \times$ speed of rotation (a being the angle rotated from A) at any position, yet the rate of cutting is proportional simply to sine a.

For, speed of cutting lines = sin a × speed of rotation,

$$= \sin a \times \text{a constant,}$$

∴ speed of cutting lines ∝ sin a.

Variation of E.M.F. during each rotation. But the induced E. M. F. set up in the wire is proportional to the speed of cutting the lines of force.

Therefore the *induced* E.M.F. *in the wire at any moment is proportional to the sine of the angle through which the wire has rotated from* A, A being the position at which the wire can only move parallel to the lines of force, and at which therefore it can have no induced E.M.F.

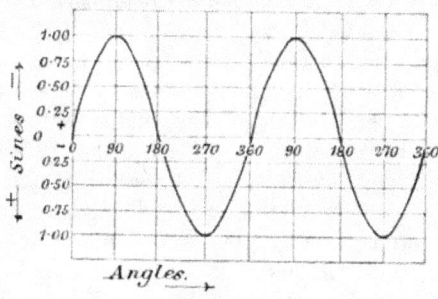

Fig. 106.

This holds strictly, both in magnitude and direction, for the sine of angles is + from 0° to 180° and − from 180° to 360°. It rises from 0° to 90°, falls from 90° to 180°, and so on.

$$\sin a = \sin (180 - a).$$

Thus in one rotation of the cylinder, the E.M.F. of induction set up in the wire varies continuously, and varies as the sine of the angle from the position A. The curve shewing the relation of the E.M.F. in the wire to the position of wire will be a sine curve.

Such a sine curve is shewn by Fig. 106. Angles of rotation are marked off as abscissae, equal distances representing equal angles. Distances proportional to sines are marked off as ordinates, positive signs being above the base and negative signs below. A sufficient number of points are marked off to enable one to draw an equable freehand curve. When drawn the curve represents the variations of the E.M.F. in the wire during rotation. The average E.M.F. will be 0·637 × the maximum, as 0·637 is the average of the sines of all angles between 0° to 90°. And one complete rotation will cause the E.M.F. to vary through a complete cycle as illustrated by the sine curve of Fig. 106.

Methods of increasing the E.M.F. It is seen that by fastening a wire to one side of a cylinder and rotating it in a magnetic field an alternating E.M.F. is set up. The magnitude of this E.M.F. at any moment is proportional to its speed of cutting across the lines of force of the field, to the strength of the field, and to the length of the wire in the field. In order therefore to increase the E.M.F. it is necessary to increase one or all of these determining factors.

Firstly, by increasing the length of the wire in the field. This may be done by fastening another wire to the cylinder at an angle of 180° from the first. Now the E.M.F. set up in this wire will be exactly equal and opposite at any moment to the E.M.F. in the first. But if one end of each of these wires be joined together the two effects will help each other. In this way the two wires will form a sort of rectangular loop, and the free ends of the loop may be joined to two rings on the axis of rotation. Now although the E.M.F.s in each wire are equal and opposite at any moment with respect to one another, yet when considered with respect to this rectangular loop they are both acting round it *in the same direction.*

Thus the E.M.F. in this loop is, at any moment, twice as great as it would have been in a single wire similarly circumstanced.

But if the wires be not fixed at 180° apart, there will be opposition of the E.M.F.s in the wires at certain positions and helping at other positions.

P. Y. 19

Again, going back to the rectangular loop, instead of being a single loop, it might be made up of a number of convolutions, the two free ends being connected to the slip rings as before. In this way the E.M.F. would be increased, for it would be the sum of the E.M.F.s in each length of wire on each side of the loop at any moment.

A longer cylinder might be used and with necessarily longer conductors fastened to it. This of course would be useless unless longer pole faces were used also, making a field of greater length *across* its lines.

The strength of the field might be increased, either by using a stronger magnet or by filling up the interpolar space with high permeability iron. The most obvious way of doing this is to make the rotating cylinder of iron; and this is adopted in practice. It must not be solid, however, for reasons explained previously; it must be built up of thin iron discs (at right angles to its axis) insulated from one another with thin paper.

The governing principles of that form of dynamo known as an *alternator* are now laid down.

Alternating Currents. The variation of the E.M.F. during one rotation of an *armature* is called one alternation. The number of alternations per minute will be equal to the number of revolutions per minute. The *frequency* of the alternations is the number of alternations per second. If the number of alternations per second is 50, for example, the frequency is expressed as 50 ∿ ; the number being prefixed by the representation of a sine curve for 360° or one cycle of E.M.F. variation. Generally the frequency (or periodicity as it is also called) chosen for practical purposes, varies between 50 and 120 alternations per second. With such frequencies the alternating current resulting may be utilised for lighting, for example, without any external evidence of the fact that it is continually varying in magnitude and direction. This will be extended later; it is merely mentioned to shew that an alternating E.M.F. may be put to practical uses.

Four-pole alternator with drum armature. In practice it is usual to have more than one loop on the armature, and to have more than two poles to the field magnets, for the number of alternations per revolution may be increased and the total E. M. F. at any moment may be made the sum of the E. M. F.s generated in the several loops.

The principle of this may be illustrated by means of a four-pole field magnet and a pair of loops, or the equivalent of four wires joined up in series and so arranged that at any moment the E. M. F.s set up in each will be acting in the same direction.

Fig. 107 illustrates diagrammatically such an arrangement. Four wires, 1, 2, 3, and 4, are shewn connected up in a series arrangement, and they are supposed to be moving together in

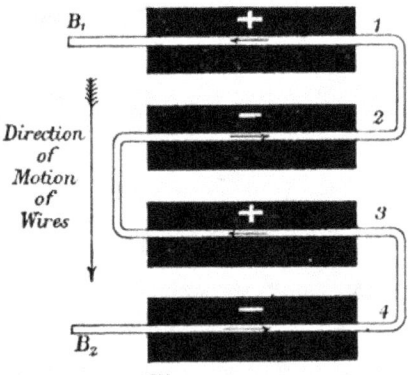

Fig. 107.

front of four equally spaced magnet poles, arranged with alternating marked and unmarked poles, the distances between the centres of each pair of poles being equal to the distance between two adjacent conductors.

With the direction of motion as shewn the E. M. F. in conductor 1 will be towards the left, of 2 towards the right, and so on. Thus all the E. M. F.s are acting along the conductors from one extremity, B_2, to the other, B_1. When this

19—2

arrangement has moved down so that all the wires are midway between the poles there will be no E.M.F. at all; and when they have moved a little further there will be a reversed E.M.F. in each conductor.

This can be put into practical form and arranged on a cylinder of iron—a *drum* it is usually called—after the manner shewn by Fig. 108, the extremities being joined up to the slip rings which make rubbing contact with the brushes B_1 and B_2.

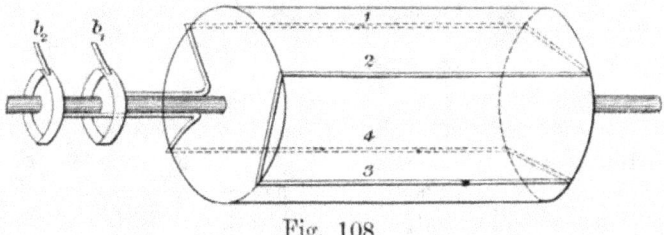

Fig. 108.

This cylinder is rotated between four pole pieces, arranged as shewn by Fig. 109, which is a diagrammatic end view of the armature and field poles.

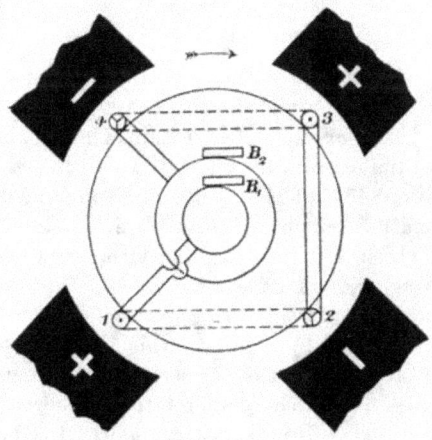

Fig. 109.

The conductors are equally spaced, 90° apart, and the connexions at the back are shewn dotted, whilst those in front are shewn in full line. The two slip rings are shewn by concentric circles, and the brushes are marked B_1 and B_2. The conductors are numbered correspondingly to the two previous figures, and the directions of the E.M.F.s in each for the particular position and direction of rotation is shewn. It is seen that they are all acting from B_2 to B_1.

With this arrangement, there will be two complete alternations of E.M.F. for one revolution of the armature, and the E.M.F. between B_1 and B_2 at any moment will be four times the E.M.F. due to one conductor.

This idea may be extended; the number of poles may be increased and the number of conductors also increased: and this is done in practically all types of alternators. Small machines, with a small output, have few poles—four will be sufficient for any output up to three kilowatts, but the large machines may have many more.

Armatures wound on a cylinder of iron, such as those described above, are known as *drum wound armatures*, and there are many different types of these, some being smooth-faced, others being slotted, and so on to meet any particular requirements.

Ring Armature. There is another type of armature known as the *ring* armature, the iron core being built up of a number of circular rings of iron, supported on the central axis by a three limbed hub or spider. This type was designed by Gramme in 1845, but it is gradually being superseded by the drum armature, chiefly because of the difficulty of winding the ring, which does not admit of lathe work or of the fixing of separately formed coils.

The general idea of a ring wound armature for an alternator is illustrated by Fig. 110, which shews a portion of the ring armature with four of the series windings and four of the pole pieces. The direction of rotation is shewn by the arrow.

The four sets of windings are equally spaced corresponding

to the spacing of the poles, and each winding is in the opposite direction to that of its immediate neighbours. The direction of the resultant E. M. F.s in each set is shewn, and it is seen

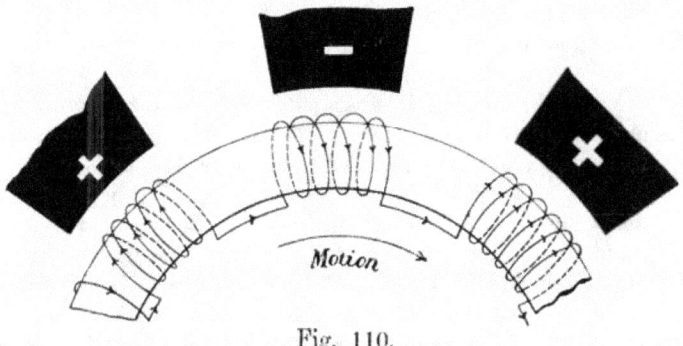

Fig. 110.

that they are all acting from one extreme to the other. The ultimate ends are connected to two slip rings as usual.

The drawbacks of the ring armature are firstly the difficulty of winding, which has been mentioned; and secondly, the fact that in each set of winding, that is in each equivalent loop, there are really two opposing E. M. F.s set up. The wires on the outside of the ring, nearer the poles, are cutting the lines at right angles and at a certain speed. But the wires on the inside side of the ring are also cutting some of the lines, and in the *same* direction though at a smaller periphery speed. Therefore in each loop there are two E. M. F.s, and the resultant is the difference between them. Of course the E. M. F. set up in the outside wires will be greater than that in the inside wires; but the fact remains that there is a tendency to waste. However it will be seen that if the iron of the ring be thick enough, the lines of force will not come through it towards the axis of rotation, but will come into it and *along* it parallel to its circumference and then out again to the next unmarked pole. In this way the E. M. F. set up in the inside winding will be zero. But the wire is waste; it is adding to the resistance of the armature, and what is equally important and disadvantageous, it is adding to the *self-inductance* of the armature.

In the drum wound armature there is less of this waste, for it is only the end connexions which are not contributing to the sum total of E.M.F. generated. There is no E.M.F. generated in these, for they move parallel to the plane of the lines of force all the time.

The field magnets of an alternator may be permanent magnets, and indeed were so in the early days of dynamo machinery; but they are electro-magnets nowadays, and are "excited" by some outside source of unidirectional E.M.F. Some alternators have an additional armature on the same shaft as the alternating current armature, arranged to yield a unidirectional current, and this is utilised to excite the field magnets. But generally the field magnets are excited from an additional continuous current generator, called an *exciter* when used for this special purpose. Field magnets are discussed in Chapter XIII.

CHAPTER XII.

CONTINUOUS CURRENT DYNAMO PRINCIPLES.

ALTERNATING currents are not suitable for all purposes. They cannot be utilised for chemical processes; and the harnessing of their magnetic effects for practical purposes involves many difficulties which are not met with when continuous or unidirectional and steady currents are used. In fact, generally speaking, at the present time the continuous current is capable of easier and wider adaptations than the alternating.

The production of a continuous current by the expenditure of mechanical energy is very similar to the production of alternating currents. The essential difference lies in a purely mechanical arrangement by means of which the E.M.F. between the brushes is kept unidirectional. This mechanical process is called *commutation*, the arrangement by which it is effected being called a *commutator*.

Simple Commutator. The idea of this may be illustrated by adapting it to a single loop, such as shewn in the discussion of alternator principles. The ends of the loop, instead of being connected to two separate slip rings, are connected to two halves of one. This ring is mounted on the axis of rotation and insulated from it. It is cut into two, each half being insulated from the other. Two brushes, B_1 and B_2 of Fig. 111, are fixed at opposite ends of a diameter of the split ring. With this arrangement the E.M.F. between

B_1 and B_2 will always be unidirectional, though not constant in magnitude; and it must be understood that the E. M. F. generated in the loop will still be alternating.

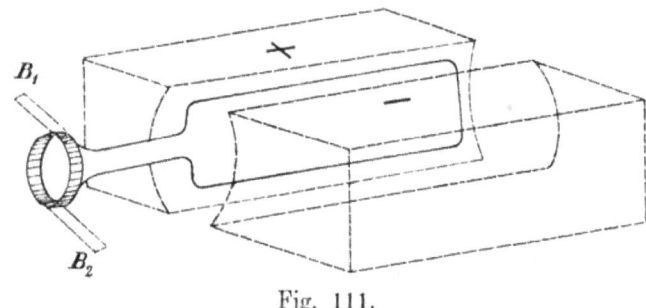

Fig. 111.

The principle of commutation is illustrated by Fig. 112, i, ii, iii, and iv. The diagrams represent an end view, looking at the commutator end of the armature, shewing the section

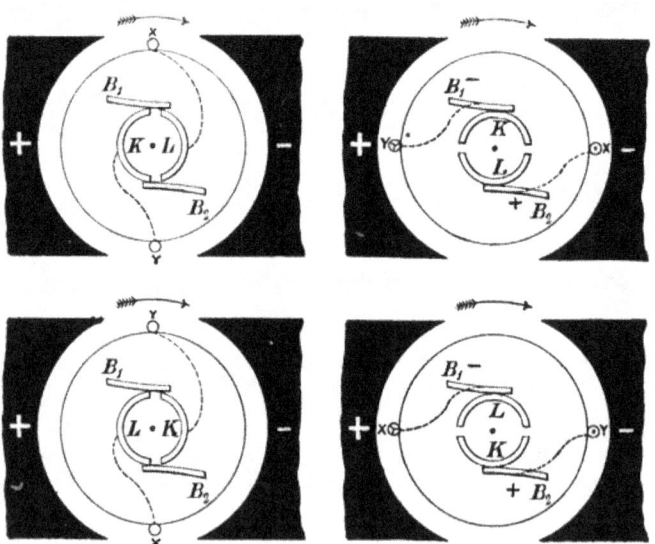

Fig. 112.

wires of the loop at X and Y. These wires are connected up
to L and K respectively, the halves of the split ring, or the
segments of the commutator. B_1 and B_2 represent the fixed
brushes, making a rubbing contact on the commutator.

Position i represents the starting position, the plane of
the loop being at right angles to the magnetic field between
the pole pieces. The direction of rotation is right-handed or
clockwise. Now at the moment illustrated by this position
the wires X and Y have no E.M.F.—they are moving parallel
to the lines of force. There is no E.M.F. between the commutator halves therefore, and at this moment the brushes are
making contact with both segments. This is done purposely,
for since the brushes *must* make contact between both segments twice during one revolution, thus short circuiting the
loop, it will be as well that they should do so at the moment
when there is no E.M.F. in the loop, and therefore no current
when it is short circuited.

The armature rotates and a moment later there will be an
E.M.F. in the loop, acting *up* the wire X and down the wire Y.
This will increase as the angle of rotation increases until after
90° have been moved through it will be a maximum. Position ii shews the armature at this moment. The E.M.F. is
acting up X and down Y, and there will be therefore an
E.M.F. between the segments L and K of the commutator
equal to the sum of the E.M.F.s along X and Y. The segment
L will be positive to the segment K at this moment. But
the brush B_2 is making contact with L; and B_1 is in contact
with K. Therefore there will be an E.M.F. between B_2 and B_1
and B_2 will be the *positive brush*.

The armature passes on through another 90°. The E.M.F.
will still be acting in the same direction, but decreasing in
magnitude until at position iii it is again zero. Again at
this position the brushes are making contact with both
halves—there is no E.M.F. in the loop and none between
the brushes.

The armature moves through the third quarter of its
revolution. The E.M.F. will now be reversed in the wires,

it will be acting *down* X and up Y, making the segment K positive to the segment L.   But now the segment K is in contact with B_2, so that B_2 *is still the positive brush* and B_1 in contact with segment L becomes the negative brush.   In this way the armature rotates through position iv, and the E.M.F. between the brushes increases between the angles 180° and 270° from the first position and decreases again to zero between the positions of 270° and 360°.

Thus it is seen that with the aid of this simple device the E.M.F. between the brushes is always unidirectional.   The armature E.M.F. is still alternating and still varying with the sine of the angle of rotation from position i.   The E.M.F. between the brushes is also varying as the sine of the angle, but with a positive sign throughout.

Fig. 113 illustrates the curve of E.M.F. between the brushes for one rotation of the simple armature described.   The abscissae represent angles, and the ordinates represent sines,

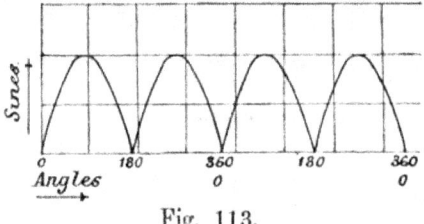

Fig. 113.

and therefore E.M.F.s for the particular angles.   There is this difference with Fig. 106, that the sines of angles between 180° and 360° have been given a positive sign instead of a negative.

There is now a unidirectional E.M.F. in the outside circuit—that is, outside the armature—but it is a variable one.

Means of increasing the E.M.F. and reducing the variation.   The magnitude of this E.M.F. at any moment may be increased in the manner previously mentioned; greater speed, a stronger field, a longer field and

armature, or a number of convolutions making up the loop.
These things will not alter its variableness, however, in ·accordance with the sine law.

But it will be seen that if two loops be arranged on this
armature so that their planes are at an angle of 90° to one
another, and if their separate E. M. F.s could be combined, the
result would be firstly a greater E. M. F., and secondly a reduced
variation. There would be no zero. This is seen by plotting
out the two E. M. F. sine curves on the same diagram and then
plotting their sum.

The full line sine curve (Fig. 114) represents the E. M. F.
variation of loop I, starting from the position where its plane
is at right angles to the magnetic field between the pole

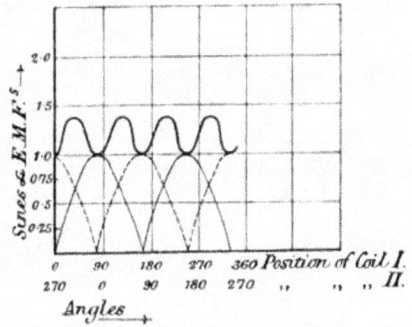

Fig. 114.

pieces. But at this position the second loop will be parallel
to the lines of force, and will therefore be in the position of
maximum E. M. F. The dotted line sine curve thus represents
the E. M. F. in loop II at corresponding positions. Thus the
two E. M. F.s may be said to be 90° apart. Now if these can
be collected and added together the resulting E. M. F. at any
moment will be their algebraic sum, and this is represented
by the thicker curve above the full line. This curve never
descends to zero, and its variation between limits is not so
great as the single loop variation.

It can be seen then that if four loops be spaced out at 45°
apart and their E. M. F.s added together, the resulting E. M. F.
will be still greater and the variation will be still less.   And
so the loops can be added until eventually a condition will be

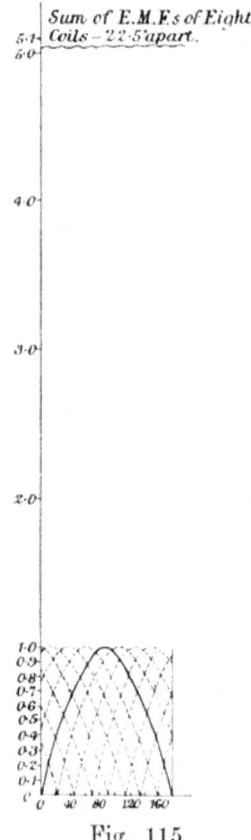

Fig. 115.

reached when the variation of the resultant E. M. F. will be
practically nothing.   Fig. 115 illustrates the effects of eight
loops, spaced at 22·5° apart; the lower portion shews the

eight equal sine curves, and the upper part shews the sum of the E. M. F.s in each loop at any moment. It is seen that the variation of this total E. M. F. is practically nothing. It amounts in this case to about 0·2 per cent.

It has now to be considered how this summation of E. M. F.s is to be arrived at practically. Clearly there must be some series arrangement in the connexions of the coils, and commutation must be strictly looked to, since in each coil there must be an alternating as well as a varying E. M. F.

There is another point to be considered. As the number of loops in series becomes greater, so does the resistance become greater. This is internal resistance, and the energy expended in it is necessarily wasted. In order to have an efficient dynamo its resistance must be very small compared with the resistance of the remaining part of the circuit in which it is to be used.

It is therefore general in continuous current generators to arrange the windings so that two or more sets of loops are joined up in parallel, each set being a number of loops in series. With two-pole machines there are two sets of loops on the armature in parallel; with four-poles there are four sets, and so on.

Winding of a Drum Armature with 16 Conductors. All these points may be well illustrated by considering the winding of a typical drum armature having eight loops, and therefore 16 conductors spaced round the drum. There are two poles to the field magnets.

Fig. 116 shews an end view looking at the commutator end. The conductors are numbered from 1 to 16 and the connexions to the commutator bars as shewn.

This commutator, instead of being a simple split ring— known as a two-part commutator—is composed of 8 segments of copper lettered from a to h, each segment being insulated from its neighbours and from the shaft on which it is built. The connexions at the back end of the drum are shewn by dotted lines.

It is assumed, for simplicity at the moment, that the magnetic field is parallel to the horizontal and at right angles to the vertical.   As a matter of fact, such would not be the case in practice because of reasons which will be discussed

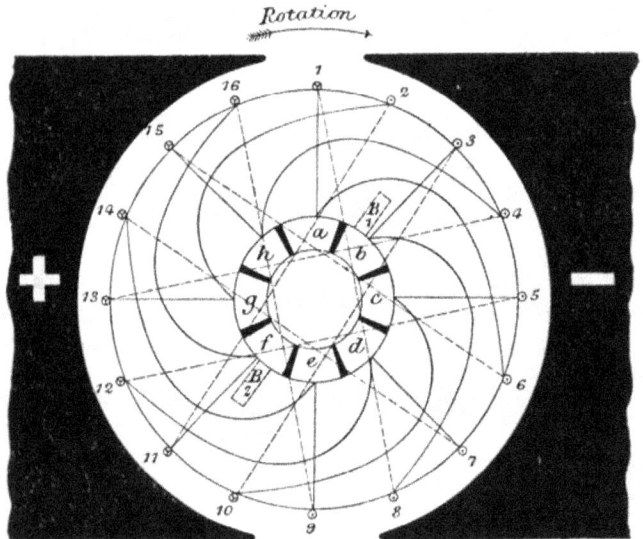

Fig. 116.

presently.   At present, then, it is assumed that the conductors cutting the vertical at right angles will have no E.M.F. set up in them, and those cutting the horizontal at right angles will have a maximum E.M.F.

The winding may be followed out by starting from the commutator bar a.   It is seen that there are two connexions to each bar; these are merely the series connexions of the coils.   The whole winding is a closed circuit of 16 conductors or 8 loops in series.

Tracing it out from a, the path of the conductors is as follows :

Commutator Bar	Conductor	Connexion across the Back	Conductor	Commutator Bar
a	1	to	8	*b*
b	3	to	10	*c*
c	5	to	12	*d*
d	7	to	14	*e*
e	9	to	16	*f*
f	11	to	2	*g*
g	13	to	4	*h*
h	15	to	6	*a*
a	1	to	8	*b*

Now when the brushes are making contact at two opposite parts of the commutator the E.M.F. between them will be the sum of the E.M.F.s of 8 *conductors* only, or of 4 *loops*, because when the brushes are in contact with bars *a* and *e*, for example, the winding becomes an arrangement of two sets of 4 loops in series, the two sets being in parallel.

This is made clear by shewing the equivalent winding in diagrammatic form, as in Fig. 117. This diagram follows the

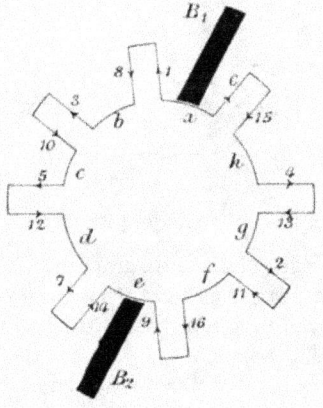

Fig. 117.

winding table exactly, the conductors and bars being numbered and lettered to correspond.

The reader should check this carefully, and work out by application of the hand-rule the direction of the E.M.F. in each conductor. He will find that the E.M.F.s in each conductor are acting as shewn on the diagram.

This is, of course, only true for the position of the armature which is shewn. But it is seen that between the bars a and e the E.M.F.s on each side are acting in relatively opposite directions—from e towards a in each half of the windings.

These E.M.F.s will be practically equal at any moment ; therefore, if there were no brushes at all on the commutator, there would be no tendency for electricity to flow round the armature coils, and therefore no waste. In fact, the arrangement is equivalent to two sets of cells, each set being made up of four cells in series and the two sets being joined up

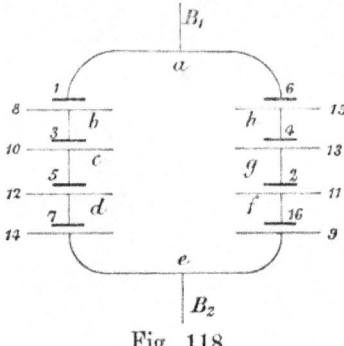

Fig. 118.

in parallel. This idea is illustrated by Fig. 118. The connecting wires may represent commutator bars, and each terminal of a cell may represent a conductor, each cell therefore representing a loop.

These have been lettered and numbered to correspond. Now the E.M.F. between a and e will be the sum of the E.M.F.s of four of the cells. In the armature, however, the

P. Y.   20

E.M.F.s are not the same in each loop at a given moment. But it will be seen that the E.M.F. in loop a 1. 8. b will be equal to that in loop e 9. 16. f; that the E.M.F.s in loops b 3. 10. c and f 11. 2. g will be equal; that they be equal also in c 5. 12. d and g 13. 4. h; and also in d 7. 14. c and h 15. 6. a.

That is to say, the sum of the E.M.F.s between a and e via the loop a 1. 8. b is equal to the sum of the E.M.F.s between a and e via the loop a 6. 15. h.

Thus, if the brushes are making contact with a and e, the E.M.F. between them will be that due to four loops, but the resistance between them will only be equivalent to that of two loops.

Now as the armature rotates the brushes will make contact with bars bf, cg, dh, and ea again in succession. But the E.M.F. between them will remain practically constant, for there will always be four loops on each side in practically similar positions, having a total E.M.F. always acting from one to the other.

But the brushes must *short-circuit* each loop twice during one revolution. Now when the armature has rotated a little further than is shewn in the figure the loop a 1. 8. b will be at right angles to the magnetic field, and will therefore have no E.M.F. set up in it. The same will apply to loop e 9. 16. f. This, then, would be a convenient and economical time to short-circuit these loops, for they are not contributing to the total E.M.F. and they are contributing to the resistance. Therefore the position of the brushes is so arranged that at this moment they are passing over the adjacent commutator segments a and b, and e and f.

As the loops are symmetrically disposed, this will apply to each loop—it will be short-circuited by a brush at the moment when its E.M.F. is zero.

This position for the brushes is found in practice by adjusting them so that there is no sparking at the brushes. If there is no E.M.F. in the coil short-circuited there will be no spark produced by the short-circuit.

It should be clearly understood that each loop of the above winding may consist of a number of convolutions, the ends of which would be joined up as shewn according to the above scheme.

Similar Ring-wound Armature.   The commutation of ring-wound armatures is much easier to follow, even though the principle is exactly the same.   Fig. 119 illustrates the winding of a ring armature which corresponds with the drum-wound armature shewn by Fig. 116.

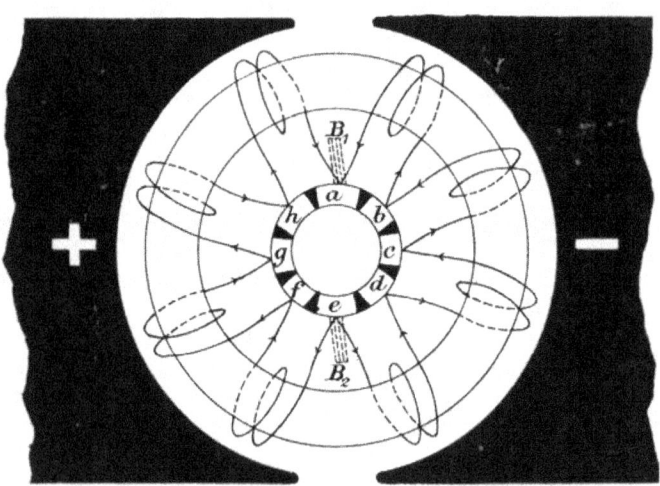

Fig. 119.

Drum Armature with an odd number of commutator segments.   There are many different ways of winding armatures, but the ultimate object and the general principle of these are identical with the method explained above.   These different methods are adopted with the view of simplifying the connexions, or of adapting the windings to some particular form of armature core, or to increase the amount of ventilation, or to allow of lathe-wound loops pre-

20—2

viously made being fixed on to the armature core directly, or to admit of proper commutation when 4 or 8 poles are to be used instead of two on the field magnets. But these various methods would fill books unto themselves, and this volume does not profess to do more than lay down fundamental

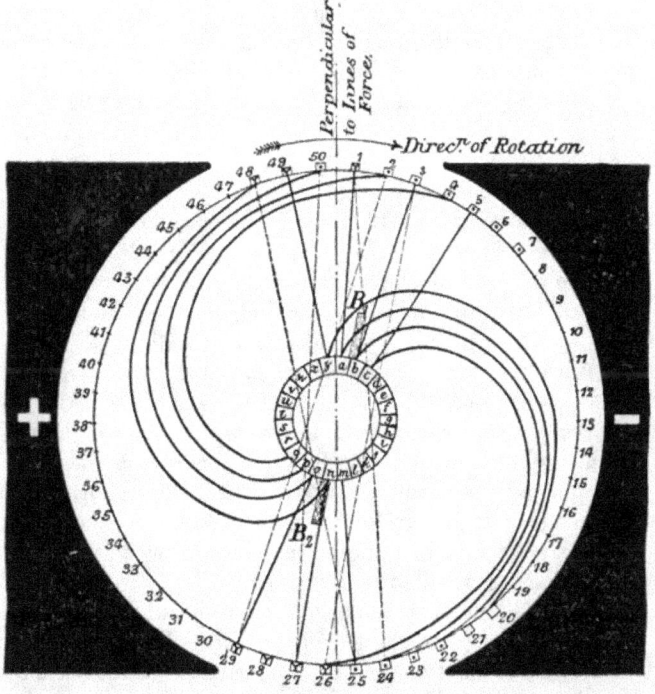

Fig. 120.

principles and shew how they are adopted to one general case; and by so doing the author hopes to have sufficiently prepared the reader that he may be able to read with profit the many excellent books which deal exhaustively with specialised branches of the great subject of Electrical Engineering.

But before dismissing the armature winding it would be

well to consider briefly the winding of a drum armature with an odd number of armature bars. These are very common in practice, and, whilst the winding may be followed out by simple extension of the winding table given for the 8-part commutator, the difference lies in the fact that there will only be one loop short-circuited at a given moment.

Fig. 120 illustrates the scheme for a 25-part commutator.

There are 25 loops and 50 conductors (or 50 × the number of convolutions on each loop). The winding table will begin :

Bar a to conductor 1 across back to conductor 24 to bar b.

Then ,, b ,, 3 ,, ,, 26 ,, c.

,, ,, c ,, 5 ,, ,, 28 ,, d,

and so on, until finally

Bar y to conductor 49 across back to conductor 22 to bar a,

and ,, a ,, 1 ,, ,, 24 ,, b.

The diagram illustrates the winding of two or three loops only in order to avoid complication, but the reader will be able to fill it up both mentally and actually.

With the odd number of commutator bars there will be 13 loops on one side of the brushes and 12 on the other for the position shewn. A moment later B_1 will be short-circuiting the loop a 1. 24. b. But B_2 will not be short-circuiting a loop at that moment, and there will be thus 12 active coils on each side of the brushes at this moment. As shewn, there will be no E.M.F. in the corresponding loops when a pair of bars are short-circuited by a brush.

This odd number works out all right when there are a goodly number ; the E.M.F.s on each side will balance off, and there will be no tendency for a current to circulate in the armature when it is running on open circuit. It can be seen, however, that this would not hold if there were only 3 loops and 3 segments to the commutator.

Calculation of E.M.F. in Armature. It has been shewn how the E.M.F. generated in a single conductor varies as the speed across the lines of force, and how this speed varies as the sine of the angle of rotation from the zero position.

During half a revolution of one conductor, then, the average
E. M. F. generated will be directly proportional to the mean of the
sines of all the angles between 0° and 180°. This mean value
is 0·637. Thus the average speed of cutting across the lines
of force is 0·637 × the actual periphery speed of the conductor,
and the average E. M. F. would be 0·637 of the maximum E. M. F.,
which would be generated when the conductor was at 90°
from the zero position.

But in the case of an armature with a large number of
conductors in series, properly commutated, the E. M. F. obtained
is practically constant. When one conductor moves from a
position its place is immediately taken by the next, and so
on, a practically constant state being maintained.

The E. M. F. generated in a wire which is cutting lines of
force varies directly as the rate of cutting. That is to say,

$$\text{E. M. F} \propto \frac{\text{number of lines cut}}{\text{seconds taken to cut them}}.$$

And
$$E \text{ absolute units} = \frac{N}{t},$$

where N represents the number of lines of force cut, and
t the time in seconds required to cut them.

$$\therefore \ E \text{ volts} = \frac{N}{10^8 \times t}.$$

In considering the total lines cut N, the length of the
conductor does not enter into the expression; for whatever
its length, so long as the same number of lines are cut by it
in the same time, the E. M. F. induced will be the same. But of
course, in a given uniform magnetic field, a longer wire will cut
a greater number N than a shorter wire. The result is that the
length is indirectly taken into account in the magnitude of N.

In the case of a single wire on a drum rotating once
round, it will cut the magnetic field between the poles *twice*.
Therefore the average E. M. F. generated in it

$$E \text{ volts} = \frac{2N}{10^8 \times t},$$

where t is the time in seconds for one revolution.

Now the speed of rotation is generally given in revolutions per minute (r. p. m.).

$$\therefore \quad \frac{r.p.m.}{60} = \text{number per second.}$$

$$\therefore \quad \frac{60}{r.p.m.} = \text{time in seconds for 1 revolution} = t.$$

∴ the average E. M. F. in the single conductor during one revolution in a field of total magnetic flux N is

$$E \text{ volts} = \frac{2N}{10^8 \times \dfrac{60}{r.p.m.}} = \frac{2N}{10^8} \times \frac{r.p.m.}{60}.$$

If there be a number of conductors round the periphery of the armature, connected in series, the average E. M. F. during one revolution will be the same in each one; and therefore the total average E. M. F. will be the sum of these.

If Z be the total number of active conductors round the armature, then of these a certain number only will be in series. In the case of the windings described $\dfrac{Z}{2}$ will be series, since the halves of the armature between the brushes are in parallel.

Therefore in the case of a 2-pole machine, with two brushes and two paths in the armature (i.e. two sets of loops in parallel), the average E. M. F.

$$E \text{ volts} = \frac{2N}{10^8} \times \frac{r.p.m.}{60} \times \frac{Z}{2}.$$

Now if there be 4 poles each conductor will cut the flux four times, and the windings will amount to 4 sets of loops in parallel. Generally, therefore, it may be written that the average E. M. F.

$$E = \frac{N}{10^8} \times \frac{p}{s} \times \frac{r.p.m.}{60} \times Z \text{ volts,}$$

where p = number of poles and s the number of parallel circuits in the armature.

It must be understood that Z represents the total number of conductors round the periphery of the armature—*not* the number of loops. If there be 16 loops, each loop having 10 convolutions, there will be $2 \times 16 \times 10 = 320$ conductors round the armature periphery. That is to say, $Z = 320$.

Distortion of Magnetic Field. When an armature is rotating and generating a current the core will be subjected to cyclic changes in its magnetisation. These will be due, firstly to the rotation between the field magnet poles, and secondly to the magnetic effect of the current in the armature windings.

In the case of a two-pole machine the armature core will be magnetised across a diameter parallel to the lines of force of the field magnets. The resulting magnetic field of the currents in the armature conductors will tend to magnetise it across a diameter at right angles to this. This resulting effect due to the currents in the various loops will have the direction of that due to those loops which are moving at right angles to the flux at the moment. The truth of this statement may be reasoned out by the reader referring back to Fig. 116, and determining the approximate direction of the magnetic field due to each loop on the armature. These fields will be equal in strength, since the current in each loop must be the same as they are series, and the direction of the resultant for each symmetrically placed pair of loops may be determined.

Position of Brushes. The ultimate effect of this is that the magnetic field between the pole pieces becomes distorted, after the manner shewn by Fig. 121. The arrow F represents the normal direction of the lines of force between the pole pieces. The arrow C represents the direction of the magnetic field due to the current in a loop round the core at the position shewn. The resultant field is shewn by the dotted lines.

Now in the previous discussions it has been assumed that the line AOB is the zero position for each loop, for at this position the loop would not be cutting any lines of force if the direction of these lines were normal. But under the conditions shewn—the working conditions—the line AOB is *not* at right angles to the resultant magnetic field—and it is this

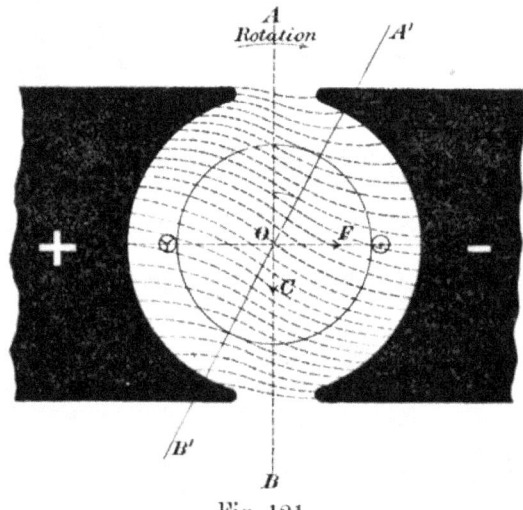

Fig. 121.

resultant distorted field which the armature conductors must cut. The line $A'OB'$ is shewn at right angles to the resultant field, and it is at this position that a loop will have no E.M.F. in it.

Therefore, instead of adjusting the brushes so that they make contact with the bars of a loop at the moment when the plane of the loop is parallel to the line AOB, they must be adjusted so that they make contact when the plane of the loop is parallel to the line $A'OB'$. The angle between these two lines, namely the angle AOA' is known as the *angle of lead*. The *lead of the brushes* is the angle through which they must be moved from a position corresponding to AOB to the proper working position corresponding to $A'OB'$. The proper position is found in practice (this has been mentioned before) by adjusting them so that there is no sparking between them and the commutator. If there is sparking it would indicate that the loops being short-circuited have E.M.F. in them at that moment.

This angle of lead is not necessarily constant for a given machine, for as the current generated by it varies, so will the magnetic effect of the current in the armature vary, and this will alter the resultant distortion of the field. At the same time the field magnets may also alter in strength and thus preserve a more or less constant angle of lead under varying loads. With a compound wound, four pole, machine there is rarely any need to alter the position of the brushes, but nevertheless all continuous current generators have their brushes so mounted that they are capable of being moved simultaneously through a sufficiently wide range to meet the remotest contingency. The arrangement for so mounting the brushes is known as a *rocker*.

The direction of the angle of lead will depend upon the direction of rotation of the armature and also upon the direction of the field between the magnet poles.

Another point which calls for comment is the fact that the effect of some of the turns will be to partly neutralise the strength of the field between the pole pieces. This is known as the demagnetising effect of the armature, and its effect is greatest with a shunt dynamo at over-load. So great is this that the resultant field becomes unstable and the machine ceases to work as a dynamo—the E.M.F. sinking back to zero. This is discussed in the next chapter.

Points on Armature Construction. It would be well here to enumerate the various points which must be considered in the building of an armature.

Drag on conductors. The drag on the conductors spaced round the periphery of an armature will always tend to cause slipping. This drag may amount to 3 or 4 lbs. per foot of conductor in those machines in which a current of some 100 amperes is passing through each. This drag is not moreover a steadily applied one—it is taken off and put on suddenly twice in each revolution, i.e. at contacts and breaks ; or, of course, a correspondingly greater number of times if there are more than two brushes. This drag together with the ordinary *Centripetal forces* due to rotation necessitate special means for fastening the conductors on to the armature core to prevent the conductors from slipping and from flying off. This may be done by having the core plates of the armature slotted all round the periphery and sinking the insulated conductors in the slots. Then when all are in place the conductors are bound round with binding wire which, by the way, runs parallel to the magnetic flux, and consequently has no induced current set up in it.

Another method may be adopted with a smooth core. The core has but a few slots—say 8—round the periphery, and in each of these an extra large conductor is sunk. Each of these large conductors is known as a *driver* and between the drivers the ordinary insulated conductors are packed. Thus every 10th conductor, say, is a driver, and these prevent slipping due to the drag. The tendency to fly off is overcome by the binding wire as before. Fig. 122 illustrates this method,

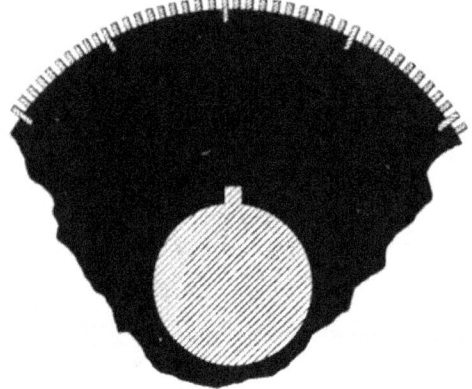

Fig. 122.

shewing a section of part of the periphery of the armature with the drivers and ordinary conductors in position.

It may be mentioned here that in large machines each loop is usually a single turn. In smaller machines each loop may be a number of turns. These loops are often formed on a lathe in the first place—bound up, and then fastened to the core.

In the large machines having single turn loops, these conductors are relatively large masses of copper having an appreciable face area on the periphery of the core. So much so indeed that eddy currents may be set up in the conductor— eddy currents, that is to say, which are merely circulating round the face area of each conductor and which are therefore waste, tending to stop the rotation of the armature. To counteract this, each conductor (this only applies of course to large conductors) is twisted through 180° laterally.

Again, because of the drag and the tendency to keep moving in a straight line, the armature core will tend to slip on the driving shaft. Therefore the core plates of the armature must be keyed on to the shaft, so that the armature may be driven without any slipping of the core. The general method of keying is shewn on Fig. 122.

Armature core. The core has to be built up of thin soft iron discs each insulated from its neighbour to prevent the circulation of eddy currents over the face of the core. The insulation is done by discs of paper of equal size to the iron discs. These latter are generally about a fortieth of an inch in thickness. The discs with their paper insulating discs are packed as tightly together as possible by means of iron end flanges which are screwed on to the shaft and fixed by means of set screws. When slotted core discs are used the slots will, of course, exactly coincide, and this result will be brought about in the stamping of the discs, for if they have all been stamped by the same machine they are bound to fit symmetrically on the shaft.

When the core has been built up the surface is insulated with special paper and with mica, which has been glued on to a thin cloth backing.

The *conductors* must be well insulated from each other and from the armature core. These conductors, in the large machines, are usually made up in the form of bars of copper of rectangular section. This is usually done by fastening a number of strips of copper together to make up each bar. These strips are soldered together at the ends and twisted laterally through 180°, and when laid on to the core, the longest part of the rectangular section lies parallel to the radius of the armature. Each bar is insulated with tape— shellac varnished—wound over a preliminary wrapping of paper. In high tension machines this insulation must be increased.

The connexions across the ends, back and front, are made with bars, or with specially formed copper strip—having an equal sectional area—but being flatter, and broader, they can be more easily adjusted to fit in together across the end. These end connexions are generally a source of difficulty in armature building, for it must be remembered that they cannot be taken straight across the back of the core, owing to the driving shaft being in the way. They must therefore be curved, and the curves must be so chosen that they will fit in symmetrically. Again, there must be some crossing at the back end, and this may be most safely done by having all the odd-numbered conductors, say, longer than the even, thus crossing the connexions at some distance apart.

At the front end the conductors are sweated on to the commutator segments.

Ventilation of armature. Sufficient cooling surface must be allowed, for not only is there heating in the conductors due to the heating effect of the current in them (and it must be remembered that this current will be a large one); but the iron core is being subjected to a rapid series of cyclic magnetisations. These cause heating of the iron core due to the hysteresis of the iron. The ventilating arrangements must be such that one minute after stopping the machine the difference in temperature between the armature core and the surrounding air must not be greater than $16.6°$ C. or $30°$ F.

This test is to be made after the machine has been running at full load for three hours.

A well designed machine can be run at 50 °/₀ overload for an hour and still answer to the temperature test.

Commutator.   The commutator is built up of a number of wedge-shaped hard-drawn copper bars, screwed round an insulating hub so that they lie radially.   They are insulated from one another and from all metal parts by means of mica, and the outer surface of all the bars forms a cylinder.

The conductors are sweated on to these bars as described. The length, width, and depth of each bar will naturally vary with the number of segments required and with the strength of the current.   In all cases a sufficient margin should be allowed for surface wear.   In many machines the commutator can be used until its diameter has decreased as much as one and a half inches.

Brushes.   The brushes were always made of fine copper gauze made up in the form of a rectangular sectioned bar with a wedge-shaped end.   They were arranged, as shewn by Fig. 123 (*a*), so that the armature rotated "with" the brushes and not "against" them.   Fine-grained carbon of

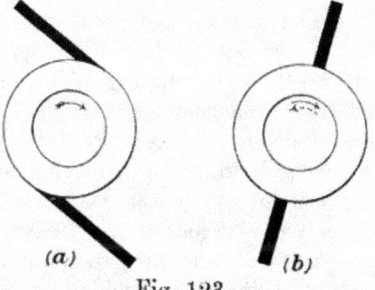

(*a*)   (*b*)

Fig. 123.

good conductivity is now used more generally, and these are set end on as shewn by Fig. 123 (*b*).   In this way the armature may be rotated in either direction.   This is especially useful in the case of motors, and it may be mentioned that carbon brushes were first introduced for these.

The gauze brushes wear away rather quickly, and if any sparking should take place small globules of copper become deposited on the commutator face, making an uneven surface, and hence increasing the tendency to spark. This cannot happen with the carbon brushes.

A spare set of brushes should always be kept in stock.

Brush holders must be designed so that the brushes are held firmly; there must be a good metallic contact to the circuit; the brushes must be capable of being fed forward automatically as they wear away; they must be held at the proper angle; and one should be able to raise them out of contact and keep them so. Moreover, the spring arrangement for pressing a brush on to the commutator surface must be capable of regulation. If the pressure be too light sparking will take place; if too heavy, ruts will be worn in the commutator. The insulation of the brushes and holders should be perfect, and they should be so designed on one framework and mounted that by means of an insulated handle they can be rocked backwards or forwards for the correct adjustment of the angle of lead.

The armature shaft. The armature must be built up on the shaft so that it is balanced—that is to say, has its weight equally distributed round the axis, with its centre of gravity at the centre of the shaft. If this is not done injurious vibration will be set up in running. The shaft must be of such material and dimensions that it can withstand the various forces upon it. These are, the weight of the armature producing bending; the driving of the armature producing torsional forces; the *magnetic pull* of the magnet for the core, which may be four or five times greater than the weight of the armature; and the lateral pull of the driving belt in the case of belt-driven machines. To overcome this it is common to have a third bearing, with the driving pulley between the second and third bearings, the armature being between the first and second. In the case of direct coupled machines this does not enter into consideration.

CHAPTER XIII.

FIELD MAGNETS.

In the first dynamos the magnetic field through the armature core was produced by means of permanent magnets. Now, however, the field is produced by electro-magnets, for in this way stronger fields may be produced for a given bulk of magnet, and moreover may be varied within limits to meet any special circumstances.

When these were first introduced, the field magnet was "excited" (i.e. magnetised) separately, by being connected up to some external source of E.M.F. such as a battery, or a smaller dynamo with permanent field magnets. These latter are known as *magneto-machines*. The separately excited machine still exists in the alternator (for an alternating current is not capable of giving permanent unidirectional magnetisation to field magnets), and in some of the large continuous current generators.

Permanent Magnets. In the case of the magneto-machine with a permanent constant magnetic field the E.M.F. generated in the armature can only be increased by increasing the speed of rotation. It can be decreased by decreasing the speed and by weakening the effective magnetic field between the poles. This may be done by shunting some of the lines of force across a piece of iron which may be placed across the limbs at any distance from the yoke. As it is removed further from the yoke towards the poles then more lines will be shunted through it, and fewer will pass across

the armature. This is the method employed in the regulation
of those small hand magneto-machines which are made and
used for the giving of " electric shocks."

Separately excited Electro-Magnets. In the separately excited machine the strength of the field between the
pole pieces may be varied by varying the strength of the
current in the magnet coils ; or by varying the number of
active turns of wire in the winding. But the field will not
be self-regulating in the least, and it is a decided advantage
to have as much self-regulation in a machine as is possible.

Self-excited Electro-Magnets. In the majority of
modern continuous current machines the field magnets are
self-excited, that is to say they are excited by a current from
the armature which revolves between their poles.

At first sight this seems to be illogical and suggests
arguing in a circle. But it must be remembered that no iron
is so soft that it loses all traces of magnetisation when
removed from the magnetising field. There will always be
some *residual* magnetisation, and this will be enough to start
the machine.

It must be admitted that in the very first instance when a
machine has just been built there may be no magnetisation in
the field magnets. Hence when it is run up to its proper
speed there can be no E.M.F. generated in the armature, and
consequently if the brushes be joined up in some way to the
field magnet windings there will be no current in them and no
resulting excitation. In such a case the magnets will have to
be excited separately at the outset, but once this has been
done, then an E.M.F. will be set up in the armature. It only
remains then to see that the direction of the resulting current
through the field windings is acting to give the same direction
of magnetisation as the initial separate current and the
machine becomes self-exciting.

When the machine is stopped there will be some residual
magnetisation, so that when it is run up again a small E.M.F.

will be set up in the armature, which will increase as the speed increases. This E.M.F. will generate a current in the field magnet, which will increase the flux through the armature. This will in turn increase the E.M.F. generated, and so on (on a sort of compound interest principle) until a steady speed and a steady state is reached.

These then are the principles of *self-excitation*. In the ordinary course of things one receives a machine from its makers, and it has been already tested, and therefore originally excited, so that one does not have to do any preliminary separate excitation. But of course it will be seen that one must start with something to work upon in a self-exciting machine.

Methods of Self-excitation of Field Magnets.

The field magnets of a self-exciting machine may be wound in different ways, so that either the whole of the current generated in the armature passes through the windings, or only a fraction; or these may be compounded.

Series Machine. A machine, the field magnets of which are wound and connected so that the whole of the current

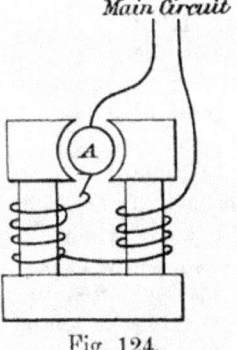

Main Circuit

Fig. 124.

passes through them, is called a *series dynamo*. The word

series signifies that the field magnet windings are connected in series with the armature and the main circuit which is being supplied from the dynamo. Fig. 124 illustrates the arrangement.

Now in this case the resistance of the field windings, together with the armature resistance, will constitute the *internal resistance of the dynamo*, and the power expended in them will represent waste.

Therefore the field windings must be of small resistance. Moreover the total current generated may be very large, hence the field windings must be of sufficiently thick wire to carry it without undue heating. Again, a large magnetic flux may be produced by a large current flowing through a few turns.

Hence it is that the field windings of a series machine consist of a few turns of very stout copper—probably copper bar or strip—having a very small resistance.

This series machine will be excited by the main current. Hence as the resistance in the main circuit decreases the current will increase, and the field excitation will therefore increase, and as a result there will be tendency for a greater E.M.F. to be generated with a resulting greater current. This is decidedly disadvantageous under certain conditions of working—when the field magnets are not magnetised up to the saturation point. It will be seen that the magnetic flux will *not increase proportionately* with the series current, because of the fact that the magnetisation of the iron will follow the form of the typical magnetisation curve, and its permeability is not a constant.

If the machine be running at a very small load one can think of the field magnets being magnetised at a point about the end of the elastic stage. The resistance is then decreased and the current consequently increased. Thus the series wound magnet becomes subjected to a greater magnetising force, and the degree of magnetisation jumps up the curve through the catastrophic stage, producing an enormous change in the magnetic flux through the armature. Up will go the

21—2

E.M.F. generated, and there is the danger of too big a rush of current in the main circuit as a result.

But with the main current of such a value that the iron is at the final stage, then it can be seen that alterations in the main resistance, and therefore in the main current, and therefore in the field magnets current, will *not* produce proportional alterations in the magnetic flux.

This is a very important fundamental consideration, and from it alone the idea forms in one's mind that a series wound machine is only a good one when it is used at an almost constant load, such that its field magnets are *just* magnetised to the degree of saturation. This is actually the case.

Behaviour of a Series Machine under a varying load. The behaviour of a machine under varying conditions of load (i.e. of E.M.F. and current generated) varies with the particular form of field excitation used. The machine is run at a constant speed, and the resistance of the main circuit is gradually decreased, corresponding readings of E.M.F. between the brushes and current strength being taken. This is continued until the output is some 50 °/₀ greater than the registered output of the machine. The results are then plotted out in the form of a curve, E.M.F. being plotted as ordinates and current strength as abscissae. The curve so obtained is known as the *characteristic curve* of the dynamo; and it has been found that all machines with similar forms of field excitation exhibit the same characteristics.

Fig. 125 illustrates the characteristic curve of a series dynamo; and it may be repeated that all series machines will give similar curves, shewing the same tendencies, but of course with different readings of E.M.F. and current depending upon the size and output of the machine. The characteristics may be plotted either as a *total characteristic* or as an *external characteristic*. In the former case the total E.M.F.s and the total currents are plotted: in the latter case the external E.M.F.s (i.e. the potential difference between the ends of the xternal circuit) and the external currents are plotted.

On the same diagram horse-power lines may be marked off
—the horse-power being E.M.F. and current ÷ 746 ; and thus
the horse-power of the machine at the various stages of the

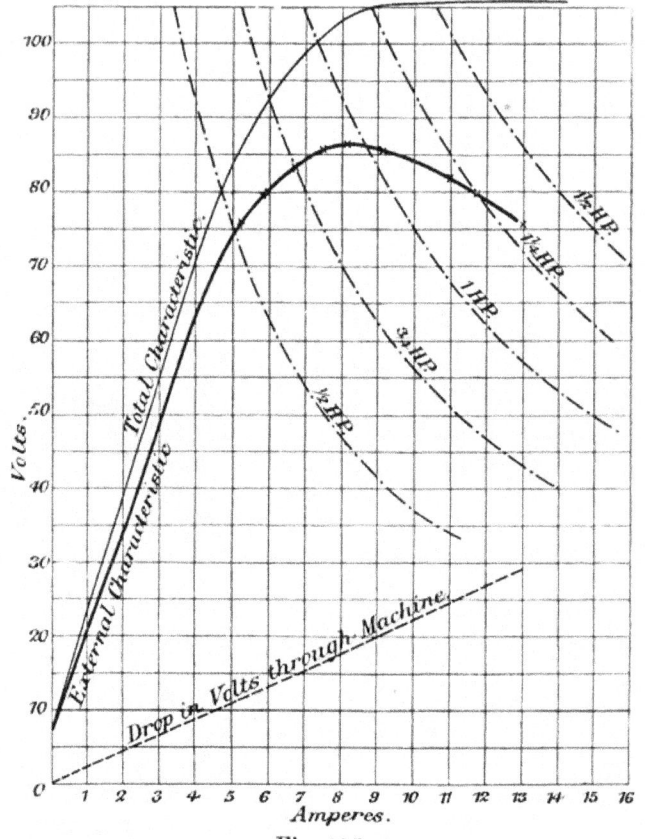

Fig. 125.

experimental test may be seen at once. These characteristic
curves so marked off enable one to see at a glance the usefulness of the machine at the various load.

The curve shewn by the diagram was plotted from results

obtained with a 1 k.w. machine. This machine was provided with different sets of spools for the field magnets, so that it could be run as a series, shunt, or compound wound machine. In this way, since all other conditions remain constant—the iron circuit, the armature, and the speed—one can compare directly the different methods of field excitation.

The following results were obtained as a series machine:

Terminal Potl. difference. Volts	Current strength. Amperes	Resis. of Dynamo. Ohms	Total E.M.F. (Current × Dyn. resis.) + Terminal E.M.F.
10	0·2	2·2	10·44
15	0·5	,,	15·1
20	1·0	,,	22·2
68	4·4	,,	77·48
75	5·2	,,	86·44
79·5	6·0	,,	92·7
83	6·8	,,	94·46
85	7·6	,,	99·72
86·5	8·2	,,	103·04
85	8·6	,,	103·9
85	9·2	,,	105
84	9·6	,,	105·12
82	11	,,	106·2
80	11·8	,,	106

The *external* characteristic is shewn by the full line, and is the relationship between the external or available E.M.F. and the current strength. The total characteristic is shewn by the thin full line. This is the relationship between the total E.M.F. and the current strength. The difference between the two curves for a given value of C represents, therefore, the drop in volts through the dynamo. This drop in volts is shewn by the straight dotted line OD. This has been plotted with (current × dynamo resis.) as ordinates and current as abscissae.

The dot and dash line curves are horse-power lines. These are drawn through points of equal horse-power. For example, the ½-horse-power line is drawn through all points at which

$$\frac{E \times C}{746} = \tfrac{1}{2}.$$

The characteristic itself shews that with a small current there is only a small E. M. F. generated in the series machine. This is a consequence of a small magnetic field being produced. Then a comparatively small increase in the current causes a big increase of E. M. F., and this corresponds with the magnetisation of iron. Then with a further increase of current there is but a small increase in E. M. F.; and a still further increase in current shews a *fall* in the external E. M. F., and the external characteristic droops.

The behaviour of the machine may be explained as follows. The magnetic field between the poles will follow the ordinary typical magnetisation curve right up to the saturation point. The speed of the armature being constant, the E. M. F. generated must also follow the same curve. Thus the total characteristic is almost an exact reproduction of a magnetisation curve. But when saturation has been reached further increases in current do not produce proportional increases in the magnetisation, and consequently do not produce proportionate increases in the total E. M. F. But since the current is increased the absorption of volts in the dynamo ($C \times$ dynamo resistance) must become greater. This will diminish the external E. M. F. and cause the droop in the external characteristic.

The matter may be compared at once with the case of the cell and the simple circuit discussed in the first chapter. The dynamo running at the stage where saturation of the field is reached has a practically constant total E. M. F. The distribution of the total E. M. F. between the internal and external parts of the circuit will be proportional to the resistance of these parts. Now if the external resistance be reduced, the current becomes greater, but the *total* E. M. F. does *not* become greater to any appreciable extent. The relationship between the internal and external resistances is now altered—the external is reduced. The total E. M. F. is practically constant; hence it follows that a greater proportion of it is now absorbed in the internal circuit than was the case before. And so it may go on, until the external resistance is actually less than the internal, in which case more than 50 °/₀ of the total

E.M.F. will be absorbed internally, and less than $50\,^{\circ}/_{\circ}$ will be measurable at the terminals of the machine. Thus it is seen that the external characteristic must droop. It is the external characteristic which one means when speaking about the characteristic curve, for the total characteristic does not give one an idea of the conditions of greatest efficiency or of the internal waste. The external characteristic gives an idea at once of the useful or available possibilities of the machine.

With the figure shewn, it is noticed that the greatest horse-power is recorded after the droop in the curve has commenced.

One cannot compare the first part of the curve to the case of a cell and a simple circuit, because the total E.M.F. of the machine is a variable one until saturation of the iron is reached.

From this discussion it would appear that a series machine is only suitable for a practically constant load, which should be such that the iron of the magnets is magnetised to saturation and the machine working at greatest efficiency.

It may just be mentioned here that the efficiency, namely, the ratio of the useful work done by it to the total work, will be found for any load by the ratio of $\dfrac{\text{external E.M.F.}}{\text{total E.M.F.}}$.

This may be obtained for any value of C by consulting the curve shewing the total and external characteristics. For example, referring to the series characteristic, when C is 8·2, the total E.M.F. is 104·5 volts, and the external is 86·5 volts; thus the efficiency is $\dfrac{86\cdot5}{104\cdot5} = 82\cdot8$ per cent.

Again, when C is 11 amps, the total E.M.F is 106·2 volts, and the external E.M.F. is 82 volts; therefore the efficiency is $\dfrac{82}{106\cdot5} = 77$ per cent.

In the case of a cell the efficiency of a cell is greatest when the external resistance is greatest. This does not hold with the series dynamo, *because there is a variable E.M.F.* But when saturation has been reached the E.M.F. becomes practically constant, and at this point the current is at the lowest value for the state of constant E.M.F., and consequently the resistance is at its greatest value. Hence at this point the efficiency must be a maximum.

Shunt-wound Field Magnets. The next method of self-excitation consists in winding the field magnets with a large number of turns of wire of comparatively high resistance and connecting the ends directly to the armature brushes. The main circuit to be supplied is also joined up to the brushes. Thus the field magnet coils are in *shunt* with the armature and the current strength in them will be the quotient

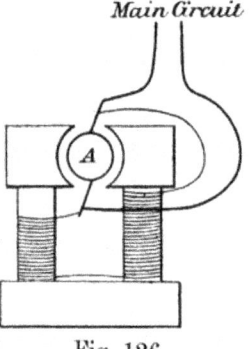

Main Circuit

Fig. 126.

of the E.M.F. at the brushes and their resistance, and will therefore be absolutely independent of the current in the main circuit, except in so far that this affects the terminal E.M.F. Fig. 126 is a diagrammatic illustration of a *shunt dynamo*.

Behaviour of Shunt Machine under varying loads. It will be seen at once that the maximum E.M.F. will be obtained at the brushes for a constant speed when there is no current in the main circuit. Of course there will be a current in the field windings, and if a characteristic be drawn between the terminal E.M.F. and the current in the field coils, it would be the same as a series characteristic. But in this case the characteristic must be drawn to shew the relationship between the main current and the terminal E.M.F. Fig. 127 is a

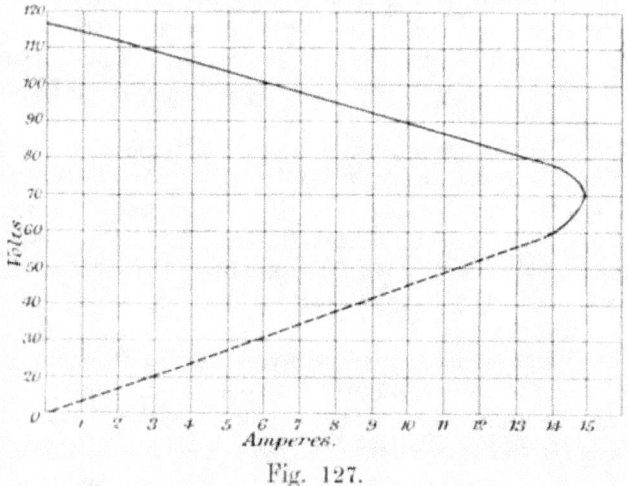

Fig. 127.

characteristic of a *shunt machine*. This is for the same machine as that used above for the series; but another pair of spools were put on to the magnet links and connected up in shunt.

The characteristic points of shunt dynamos are shewn by the figure. With a zero external current the E.M.F. at the brushes is a maximum, when the machine has been run up to its normal speed. But as soon as a resistance is put in the main circuit, and some current passes through it, the E.M.F. at the terminals drops. Now it must be remembered that the main circuit and the field coils are in parallel. Hence, whatever resistance is put in the main circuit, the joint resistance of the two paralleled circuits must be less than the smaller of the two. Hence, the total resistance being less, the current in the armature will be greater; and therefore the drop in volts through it will be greater. Therefore the E.M.F. at the brushes will fall, and then the current in the shunt coils will fall, and so on. Further, as the current in the armature increases, the demagnetising effect of some of the coils becomes greater, whilst the shunt-excited field is becoming slightly smaller. This will go on as the main circuit current increases (denoting a decrease in external resistance), until the magnetisation of the field becomes unstable, and the E.M.F. generated falls away to zero, the current falling, of course, with it.

Thus there is a gradual decrease in terminal E.M.F. from the outset, and when a certain load has been reached it sinks away to zero quite suddenly. In this case the magnetisation starts well beyond the saturation point, remains practically saturated even with a falling current in the field coils, then starts on the downward part of the curve; and here the trouble begins.

Shunt machines have not a very great range of usefulness. The great point about them is that as the current in the main circuit decreases the E.M.F. *increases*. And this is exactly the reverse with the series machines. The special sphere of usefulness for the shunt machine is the charging of secondary cells, where a back E.M.F. is set up which decreases the current. Now if the E.M.F. of a charging machine dropped so that it became lower than the back E.M.F. of the cells, it would start *motoring*. That is to say, it would be driven by the cells which would be thus running themselves down. The E.M.F. of a series machine will drop if the speed of running is reduced, *or* if the main current is reduced very much. A shunt machine will only have its E.M.F. reduced by decreasing the speed or by *increasing* the main current. Assuming, therefore, that the speed remains constant, the E.M.F. of a shunt machine will increase as the back E.M.F. of the cells increases, and so decreases the strength of the charging current; and thus there is no fear of the machine being driven as a motor by the cells. By the way, it may be stated that the machine would run in the *same* direction when motoring as when being driven; but the current in its armature would be reversed.

Compound-wound Field Magnets. The most general method of self-excitation consists in combining series and and shunt windings on the same machine. In this way the advantages of both are obtained, whilst their disadvantages are more or less overcome, since the disadvantages of each method shews itself at the beginning and end of the range of output. A machine with combined winding is known as a *compound machine*. Fig. 128 shews the diagrammatic

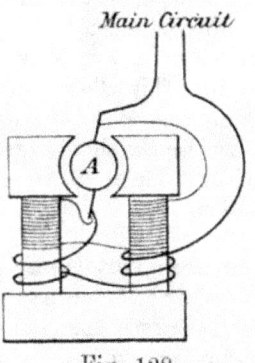

Fig. 128.

scheme, and it seems that the field is partly excited by a few series turns, in series with main circuit and armature;

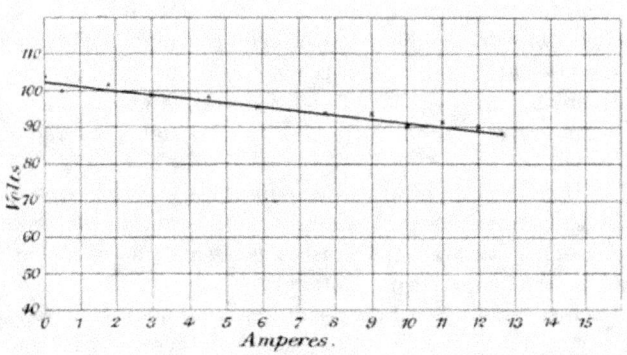

Fig. 129.

and partly by a set of shunt windings in parallel with the main circuit.

Fig. 129 is a characteristic curve of a compound machine; the same machine as before, but with compound-wound spools on the field magnets. It is seen the E.M.F. is practically constant over a considerable range of current variation. The "compounding" will only hold with a given machine with the registered speed of rotation. At other speeds, and consequent E.M.F.s, the series and shunt windings will not balance off one another's defects.

It is customary to have a variable resistance in series with the shunt field coils. This allows one to vary the current strength in the shunt coils, and, by cutting out some of the shunt resistance, increase the shunt current and the resulting magnetisation, and thus maintain a constant E.M.F. against the armature reactions.

Separate-armature excitation. Another method of self-excitation, and one which can be used for alternators or continuous current generators, consists in having an additional commutated armature on the same shaft and between the same poles as the armature for supply of the main current. The brushes of this additional armature are connected to the ends of the field windings, which of course will consist of many turns of comparatively high resistance. With a given speed there will be therefore a perfectly constant field magnetisation—but the E.M.F. between the brushes will vary in accordance with the simple relationship between the internal resistance of the armature and the resistance of the main circuit.

It is usual to have in a case like this a variable resistance in series with the field winding. This can be cut down as the E.M.F. between the brushes is seen to drop, and it can then be restored to its initial value.

Instead of a separate armature, some additional windings may be put on the main armature, and these may be commutated at the other end of the shaft away from the main

commutator; or in the case of an alternator from the slip
rings. This is the general form of excitation of alternator
field magnets.

In addition to the methods mentioned there are many
various modifications, such as partly by series and partly by
separately supplied windings. However it is not expedient
that these should be discussed in this volume.

Forms of field magnets. In designing a field magnet
the main question to be considered is the magnetic circuit.
The magnetic conductivity of the path of the lines of force
should be as good as possible. To that end the area of cross-
section of the iron should be sufficient for the maximum flux,
and this area should be such that there is at least one square
centimetre for every 16,000 lines of force. This of course
varies with the quality of the iron used.

The length too should be kept down, and the same applies
to the number of joints of the various parts—limbs, yoke and
pole pieces. In short, the best magnetic circuit will be the
shortest, having the greatest area of cross-section, and made
with a material of highest permeability and with a minimum
of joints. Of course it must be remembered that the greatest
reluctance must be offered at the air gaps between the pole
pieces and the armature core, so that a joint more or less will
not alter the total reluctance to any great extent. But at
the same time a designer cannot be wrong in keeping down
the reluctance as much as possible.

However, there are other things to be considered than the
ideal magnetic circuit. There is the cost and the mechanical
part which the magnet may have to play in the construction
of the machine. All the items must be considered together—
but the fundamental thing must be the magnetic circuit of
greatest conductivity and least waste for a given cost and to
fulfil given mechanical conditions.

With these equally fulfilled there is no special virtue in
one form of field magnet rather than another.

Fig. 130 illustrates diagrammatically some typical forms of

2-pole field magnets. Nos. 1 and 2 are similar in general form. No. 1 is an old type and No. 2 is an improvement upon it, by being greater in cross-sectional area and shorter

Fig. 130.

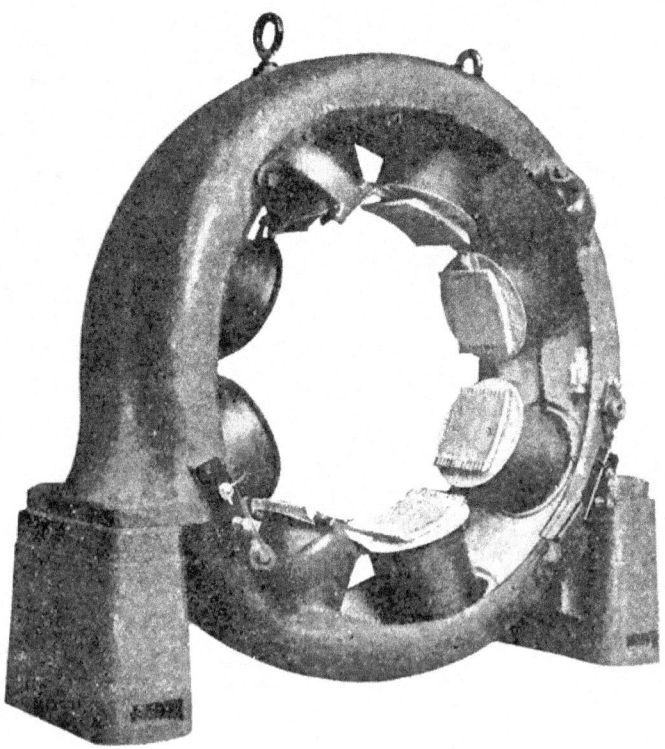

Fig. 131.

in length. The yoke and pole pieces are bolted on in both cases, making four joints in the circuits. The special difficulty about this "under-type" lies in the fixing to the bedplate. Obviously this must *not* be of iron or the greater part of the flux would be shunted through it. No. 3 is known as an "over-type." No. 4 represents a field magnet with a double circuit having consequent poles at the pole pieces. This amounts to a ring magnetised across a diameter.

The points to be noted are the tendency for leakage of lines of force—that is for lines of force to pass into the air without also going through the armature core; and that all cross-sections of the circuit are sufficient to carry the maximum flux.

It is general nowadays to have enclosed forms of machines, and the multipolar types of field magnets are responsible for this being specially adopted.

Fig. 131 illustrates the field magnets of an 8-pole machine. It is from a photograph and shews the exciting spools and the faces of the poles. Other multipolar machines follow the same form. In fact this is the general form for yokes of all machines of more than 20—30 kilowatts output.

Pole shoes. It is usual in continuous current machines to spread out the face of each pole so that it covers a greater area than the cross-section of its core. This is done to reduce the reluctance, or magnetic resistance of the air gap between the magnet pole and the armature core, by increasing its equivalent area of cross-section. At the same time however it must be remembered that the polar flux density will in this way be reduced, and the pole face—or *pole-shoe*—may not be saturated. This has been found to be disadvantageous, and the difficulty is overcome by making the pole face of less permeable material than the core.

Lamination of pole pieces. With some forms of armature with slotted cores, it must follow that as the core rotates the movement of each tooth over a pole face must tend to alter the distribution of lines of force. This will be continually taking place and hence eddy currents are likely to be set up in the pole face. To prevent these currents pole cores are frequently laminated and insulated. In other cases, the pole shoes only are laminated in this way because the eddy currents will not be set up to any great depth in the iron, since the magnetic variation will only take place near the pole surface.

Magnetic circuit of a dynamo. The general principles of a magnetic circuit have already been discussed at some length in Chapters VIII. and IX. It was there shewn how to determine the ampere-turns necessary to give a certain total magnetic flux N through a circuit, and examples were given of the effect of an air gap.

However, no account was there taken (in a quantitative sense) of *leakage*. Now in every magnetic circuit of a dynamo or motor there will be magnetic leakage—stray lines of force which are not passing through the armature core as desired. So that of the total flux generated by a given number of ampere-turns, a portion is wasted and the remainder represents useful magnetic flux.

The ratio of the leakage flux to the useful flux is called the *dispersion*, but the more generally used ratio is that of the total flux to the useful flux. This is called the *coefficient of dispersion*, and it is generally denoted by v.

$$\text{Coefficient of dispersion } v = \frac{\text{total flux}}{\text{useful flux}},$$

$$= \frac{\text{total induction } N}{\text{no. of lines of induction passing through armature}},$$

$$v = \frac{N}{N_a},$$

where N_a represents the lines in the armature.

This coefficient varies, but it will always be greater than unity, and in the ordinary machine it rarely exceeds 1·75. It also varies with a given machine, for the leakage will depend largely upon the degree of magnetisation of the field magnets. There will be greater leakage above the saturation stage than below it, and armature reactions tend to increase the dispersion by the very fact of their distorting the field between the magnet poles.

The coefficient of dispersion may be determined by the experiment described on page 264. By calculating N at the yoke and at the armature the ratio of one to the other may be determined. This may be done for different values of the magnetising current in the magnet windings and it will be seen that the coefficient increases with the excitation. The same scheme exactly may be applied to a dynamo. A "search coil" of one or two turns may be wound on the armature—at right angles to the flux—and one of equal turns on the yoke. The Ballistic kicks obtained when each is connected to a Ballistic galvanometer in turn and a current is "broken" in the field coils, will give a measure of the flux in each coil. The ratio of the kicks will give the required coefficient of dispersion.

Having determined a coefficient of dispersion it follows that

$$\text{The flux across the armature} = \frac{\text{total flux}}{\text{coefficient of dispersion}}.$$

If the coefficient of v is known, it follows that the total Magnetic Motive Force ($1\cdot25SC$) necessary to produce a given induction N through the armature core will be given by

$$\text{M.M.F.} = (N_{\text{arm.}} \times \text{Reluctance}_{\text{arm.}}) + (N_{\text{gap}} \times \text{Reluctance}_{\text{gap}})$$
$$+ (N_{\text{limbs}} \times \text{Reluctance}_{\text{limbs}} \times v).$$

It is possible to calculate the probable dispersion, and rules have been drawn up enabling one to do this. These do not come into the scope of this work.

It has been shewn that the total flux $= \dfrac{\text{total M.M.F.}}{\text{total reluctance}}.$

Hence in field magnet calculations one has to determine the

reluctance of each part of the magnetic circuit—the yoke, the limbs, the armature core, and the air gap.

The chief difficulties lie in the determination of the *mean length* of the magnetic circuit and the *effective area* of cross-section of the iron.

This is largely a matter of discretion and of appreciation of fundamental magnetic principles.

CHAPTER XIV.

MOTORS.

An electric motor is the precise converse of a dynamo. It is a machine by means of which electrical energy may be converted to mechanical energy. A dynamo may be uncoupled from its driving engine and a current of electricity be supplied to it, in which case it will rotate as an electric motor. Indeed it may be said that on general lines the construction of a motor is identical with that of a dynamo. One machine may be used for either purpose.

The fundamental principle of the electric motor was laid down at the beginning of Ch. XI., and the reader is referred back to that introductory discussion. When a current is urged in one direction through a wire, that wire tends to move across the magnetic field. When the wire is moved across the magnetic field in the same direction a current tends to be set up in it in the opposite direction.

When a dynamo is running and generating electricity there is a current flowing in the armature conductors. This current tends to drag the conductors in the opposite direction to the direction of drive. In short the dynamo *tends* to be a motor—running in the opposite direction—and this tendency increases as the output of the dynamo increases. For this reason, it may be said, greater mechanical forces will be necessary to drive the dynamo at the same original speed. In a dynamo the mechanical drag on the conductors and armature (that is the *motor drag*) is in opposition to direction of drive. This, of course, is inevitable according to the principles of the conservation of energy. In the dynamo it may be said that the work done in overcoming the opposition drag—the motor drag—will be the source of the energy of the currents produced.

Motor Commutation. Thus it is seen that a dynamo is always endeavouring to run as a motor, and a motor is always endeavouring to generate current as a dynamo. It will be assumed that dynamos and motors are constructionally the same so far as continuous current machines are concerned. The same necessity for commutation will be at once apparent. Each conductor must have its current reversed as it passes from the sphere of action of one pole of the field magnet to that of the next—an opposite pole—in order to maintain the same direction of mechanical drag. Thus the commutation will be identical with that of the dynamo.

The armature construction will also be similar. There will be the same, if not a greater, necessity for the lamination and insulation of the iron parts. There will be the same forces to consider in the fixing of the conductors on the armature core, and in the designing of the core generally.

There will be a difference in the position of the brushes however. Firstly, if it be considered that a dynamo is being run as a motor in the *same* direction as that in which it was driven as a dynamo, it is obvious that the currents in the armature coils will be relatively in opposite directions, other things being the same. Therefore it follows that the distortion of the magnetic field due to motor armature will be in the opposition direction from the zero position to that created by the dynamo armature.

Fig. 132a illustrates the distortion of a dynamo armature, the angle AOB being the angle of lead and the line BB' the resultant "zero position" or line at right angles to the resultant magnetic field. The figure b shews the same machine as a motor under exactly equal conditions. The new distortion is shewn, and the line BB' is the line at right angles to this resultant field.

Now it must follow that as each loop becomes parallel to *BB'* the mechanical drag on it will be zero. This then will be a favourable moment to short-circuit it. Thus the brushes would be put in such a position that this was accomplished.

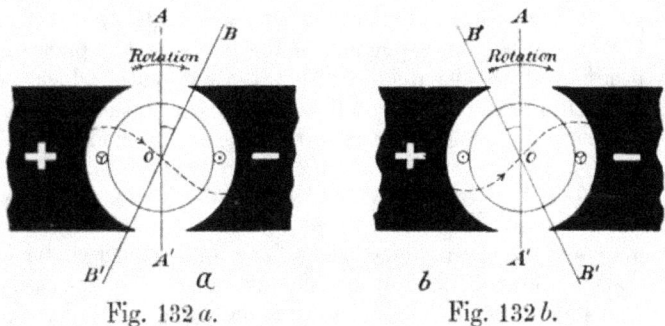

Fig. 132 *a*.   Fig. 132 *b*.

The angle *AOB'* is the angle of lead—but whilst it is equal in magnitude to the corresponding dynamo lead it is in a different direction. It is called a *backward lead*, whilst the dynamo brushes are set at a *forward lead*.

Back E.M.F. But there is another reason for this position of the brushes, and it lies in the fact that whilst a motor is rotating each loop is cutting lines of force, and that as a result there must be an E.M.F. of induction set up in each loop. Now this E.M.F. must be in opposition to the E.M.F. which is urging the current necessary to drive the motor. This follows from the previous discussions, and consequently it is called a *Back*—or *Counter*—E.M.F. Further, the Back E.M.F. in each loop will vary in accordance with the sine law previously established, and consequently will be zero in each as it becomes parallel to the line *BB'*, which is at right angles to the resultant magnetic flux. At this moment then each loop is short-circuited by the brushes.

Mechanical Drag on each Conductor. The drag on each conductor spaced round the armature will depend upon the strength of the current in it, upon the strength of the magnetic field between the poles, and upon the length of conductor in that field. Each conductor will tend to be dragged perpendicularly across the field, but for each position around the armature core this tendency may be resolved into two, one along a radius and one at right angles to it, acting, that is to say, at a tangent to the circle at the particular position of the conductor. The resolved parts, which are useful in producing the rotation, are those components acting tangentially to the periphery of the core at any moment.

Torque. The tendency of this force to produce a turning motion is called the *turning moment* or the *torque*, and it is measured by the product of the force and the perpendicular distance through which it acts at any instant. In the case of a single conductor the torque will be measured by the product of the tangential force and the radius of the circle in which it rotates.

The total torque or the *torque of the motor armature* will be the *sum* of the torques of each conductor at a given instant.

The mechanical drag on a single conductor in a magnetic field tending to pull the conductor across the field at right angles

$$= H \times \frac{C}{10} \times l,$$

where H is the strength of the field, C the current in the conductor in amperes, and l the length of the conductor acted upon. This assumes a straight conductor and a uniform field. The drag will be given in *dynes* when H is in c.g.s. units, C in amps, and l in centimetres.

This will give the force acting on a conductor of a motor armature at the moment when it has rotated 90° from the zero position.

Effect of Back E.M.F. in a Motor. It has been pointed out that when a motor is running there is back E.M.F. generated because of the fact that the armature conductors are

cutting lines of force. It follows therefore that the current
which is passing through the motor will *not* follow the simple
Ohm's law value and be equal to the applied E.M.F. ÷ the
resistance.

The E.M.F. which is actually being used to maintain the
current in the motor will be the difference between the applied
E.M.F. and the back E.M.F.; as in the case of a polarised voltaic
cell. Thus

$$\text{Current strength in motor} = \frac{\text{applied E.M.F.} - \text{back E.M.F.}}{\text{motor resistance}},$$

or
$$C = \frac{E - E'}{R},$$

where C is the motor current, E the applied, and E' the back
E.M.F.s; and R the motor resistance.

It is clear therefore that with a given applied E.M.F. a
certain motor will have varying currents in it at varying
speeds, because the back E.M.F. will vary with the speed of
rotation of the armature.

If the armature be prevented from rotating the current
would follow Ohm's law exactly. The current would be large
—too large indeed for the motor, if the normal registered
applied E.M.F. be used. The reader is warned not to attempt
the experiment. But the resistance between the motor terminals may be measured (by some fall of potential method since
it will be low) and may be denoted by R. The current in a
stationary motor, $C = \frac{E}{R}$.

Now as the motor runs up this current must gradually
decrease, since the back E.M.F., E', will gradually increase.
Further, with a given applied E.M.F. the current in a given
motor must be a *maximum* when the armature is kept
stationary, for at any speed there is bound to be some back
E.M.F., and $E - E'$ must always be less than E.

Let the motor be run up to its normal registered speed
(i.e. the speed it is designed for), and under these conditions let

$$C = \frac{E - E'}{R}.$$

The watts supplied to the motor $= E \times C$...........(1).

The watts expended internally in the motor will be wasted. This waste will be represented by the product of the current in the motor and the drop in volts through it.

The drop in volts through the motor must

$$= R \times C.$$

But
$$C = \frac{E - E'}{R}.$$

$$\therefore R \times C = R \times \frac{E - E'}{R} = E - E'.$$

$$\therefore \text{ internal watts wasted} = (E - E')\, C \,........(2).$$

The difference between the watts supplied and the watts wasted will be the watts transformed to mechanical energy. That is, the difference between expression (1) and expression (2).

Therefore watts transformed

$$= (E \times C) - \{(E - E')\, C\}$$
$$= EC - EC + E'C = E'C ....(3).$$

The total watts must be the sum of the watts transformed and the watts wasted internally.

Therefore expression $(1) = (2) + (3)$.

Now the "watts transformed" must be regarded as being made up as follows, since expression (2) merely signifies *internal* watts wasted.

Watts Transformed

Hysteresis : Eddy Currents : Friction	Useful mechanical work done by motor
Losses	
h	w

Clearly the *mechanical efficiency* of the motor will be $\dfrac{w}{EC}$.

The *electrical efficiency* is the ratio of $\dfrac{\text{watts transformed}}{\text{total watts}}$

$$= \frac{h + w}{EC} = \frac{E'C}{EC} = \frac{E'}{E}.$$

Efficiencies have been expressed as a ratio of *watts*; but it will be seen that watts represent a definite quantity of *work per second*, and therefore the expressions have a proper significance.

This holds good always, but of course the back E.M.F., E', will vary with different loads. As the motor is called upon to do more work it will tend to slow down, and the back E.M.F. will drop. A larger current will then pass, and the speed will go up again in a properly governed motor. But the internal watts must be greater, and consequently the external watts will be a smaller proportion of the whole, and the value of E' will diminish. Indeed, if the mechanical load on the pulley be great enough, the motor will be pulled up and E' will be zero.

When the machine is running without external load E' will be more nearly equal to E than under any other circumstances. In the ideal perfect motor E' should equal E under these conditions. That would mean that there would be no friction at the bearings of the machine, no hysteresis and eddy current losses—an impossible thing in practice.

The back E.M.F. cannot be measured directly. A voltmeter connected to the terminals of a running motor will only indicate the applied E.M.F., E. But if the current passing be measured by means of an ammeter connected up in series with the motor the back E.M.F., E', may be calculated if the motor resistance R is known. For since

$$C = \frac{E - E'}{R},$$

$$\therefore \ E' = E - (C \times R).$$

Determination of Mechanical Efficiency of motor by brake test. The efficiency of a motor may be determined by what is known as a *brake test*. This consists in measuring the mechanical energy obtained from a motor by overcoming a certain brake force applied at the circumference of the pulley, and determining the ratio of that to the electrical energy supplied in the same time.

The "brake" may be rigged up as shewn by Fig. 133. A length of webbing is arranged as shewn under the pulley, and one end is fastened to the hook of a spring balance, whilst the other end is fastened to a cord which passes over a single pulley, terminating at a hook on which weights may be hung.   By altering these weights the brake force may be

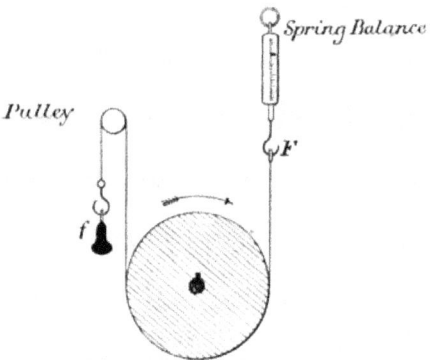

Fig. 133.

altered to any extent.   When the motor rotates it will tend to drag the webbing down on the side of balance.   The tendency will be overcome by the reaction of the spring balance, and the webbing will slip on the pulley.   If F repre-sents the force indicated by the balance, and f the weights on the hook, the brake force applied to the pulley by the slipping belt must be $(F-f)$.

Care must be taken in starting, and to that end the weights f should be held off by the hand and allowed to operate gradually.   Otherwise the friction between belt and pulley might be great enough to cause too heavy a drag on the balance with possible disastrous consequences.

Now the work done on the brake in a given time will be equal to the product of the brake force and the distance through which it was overcome.   Thus in each revolution the work done will be the brake force × the circumference of the pulley, in proper units.

The following results must be determined :

Circumference of pulley	$= d$ cms.
Reading of balance	$= F$ grammes.
Weight on hook	$= f$ grammes.
$\therefore$ Brake force applied	$= (F-f)$ grammes.
Number of revolutions per minute	$= n$.
Applied E.M.F.	$= E$ volts.
Current supplied	$= C$ amperes.

The work done on the brake in 1 minute will be

$$(F-f) \times d \times n \text{ gramme-centimetres}$$
$$= (F-f) \times d \times n \times 981 \text{ } dyne \text{ } cms. = \text{ERGS.}$$

In the same time (60 seconds) the electrical energy supplied to the motor

$$= E \times C \times 60 \text{ } joules.$$

Now 1 joule $= 10,000,000$ ergs ; and to get the efficiency the external work must be expressed in the same units as total work.

Hence the brake work, i.e. external work,

$$= \frac{(F-f) \times d \times n \times 981}{10^7} \text{ joules.}$$

Therefore the mechanical efficiency of the motor

$$= \frac{(F-f) \times d \times n \times 981}{10^7 \times E \times C \times 60} .$$

The efficiency should be determined for different values of the brake load, and a curve may be plotted shewing the relationship between the efficiency and the load. It will be found generally that a motor has its greatest efficiency when it is working at the maximum load for which it was designed. Thus a 20 k.w. motor will be more efficient when it is working at 20 k.w. rate than at 10 k.w. rate or at a higher rate. A good motor may be run at an overload for some time ; but it will be less efficient. As in the case of all machines, a maximum rate of working is not a maximum of efficiency.

Brake Horse-power. The brake horse-power of the motor may be determined from the above experiment.

The work done in one minute

$$= \frac{(F-f) \times d \times n \times 981}{10^7} \text{ joules.}$$

∴ in one second is

$$\frac{(F-f) \times d \times n \times 981}{10^7 \times 60}.$$

The rate of working in joules per sec. = watts, and 746 watts = 1 horse-power.

Hence the brake horse-power

$$= \frac{(F-f) \times d \times n \times 981}{10^7 \times 60 \times 746}.$$

The horse-power supplied $= \frac{E \times C}{746}$.

If, in the above experiment, F and f be in *pounds* and d be in inches.

The work done in one minute on the brake will be

$$= \frac{(F-f) \times d \times n}{12} \text{ foot-pounds.}$$

33,000 foot-pounds per minute is the rate of working known as 1 horse-power.

Therefore the brake horse-power

$$= \frac{(F-f) \times d \times n}{12 \times 33,000}.$$

It may be mentioned that 1 foot-pound of work

$$= 13\cdot56 \text{ joules.}$$

Motor Starter: no-load release. It is necessary that all electric motors should be provided with a *starter*, that is, an easily adjusted variable resistance placed in series with the armature. If a motor be connected up to the source of supply of E.M.F. and the switch be put on, there will be a big rush of current in the armature, because it has a low resistance, and being stationary there is no back E.M.F. This rush would, in nearly all cases, be too much for the armature conductors and would be liable to produce damage. Moreover, it would be wasteful. Therefore it is necessary to have some resistance in series with the armature when the current is first switched on. The armature will rotate with an acceleration. When it becomes steady a little resistance is cut out; the armature will go quicker. Then more resistance is cut out, and so on, until all the resistance has been cut out, and the motor is running at free normal speed.

The *starters* in general use are also provided with a *no-load release*, by means of which the resistance is thrown in again when the current is switched off. In this way no damage can be done by an operator switching off and then switching on again without having previously put in the resistance.

The general arrangement is shewn by Fig. 134. A contact arm is capable of motion about P, and can make contact with a series of studs numbered 6, 5, 4, &c. Between each pair of studs there is a resistance coil, and between 0 and 6 they are all in series connexion. The contact arm is held by a spring against a stop S, and a piece of iron I is fixed on one side in such a way that when the arm is swung round against the spring to the stud 0, it comes into magnetic contact with the poles MM' of a little electromagnet placed as shewn.

The starter is shewn connected for use with a shunt motor. When the main D.P. switch is on a current will pass around the field magnet coils and also through the small electro-magnet MM'. But no current can pass through the armature until the arm is moved to stud 6. Then a current will pass; but the armature will be in series with the six sets of coils. As the arm is moved round the resistance in series is cut out in steps until at the stud 0 the armature will be connected virtually to the positive terminal of the supply. Now at this position the electromagnet MM' will hold the arm against the spring, and the motor will run normally.

When the D.P. switch is broken, the MM' will cease to be

magnetised, and the arm will fly back against the stop *S*. This is the *no-load release* : for now, when the D.P. switch is put on again, the motor can only be started by again moving the arm round.

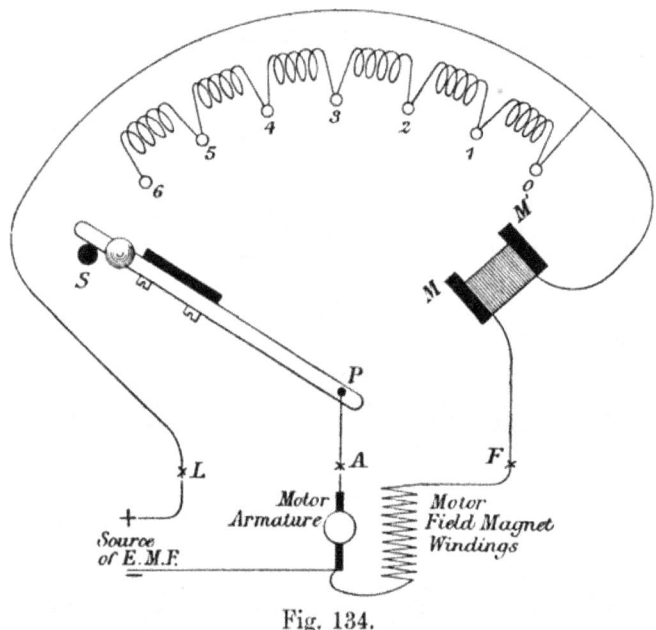

Fig. 134.

Some starters are also provided with an *over-load* release which merely consists of a small electromagnet in *series* with the *armature*, and arranged to pull up a piece of iron when the current becomes greater than the normal working current. This piece of iron, when pulled up, connects two studs together, the studs being connected to the ends of the electro-magnet coil *MM'*.   Thus the coil becomes short-circuited, and the arm will be released.

The drawback to this starter lies in the fact that there is nothing to prevent a wilful or ignorant person from quickly pulling the contact arm right across to stud 0.   If he does so the whole object of the starter is defeated, and damage may be done.   There are some forms which prevent such a possibility.

Field excitation of Motors.   The field magnets of electric motors may be wound in series with the armature, or in shunt.

The series motor has certain distinct advantages over the shunt, the chief of which is that when the current is switched on to the stationary motor there will be a large *starting torque,* as the strength of the current in a motor must be greatest when there is no back E.M.F.   But the current in the field magnets will be the same as that in the armature, and the torque will be proportional to the product of the field magnetisation and the armature current.   Thus it is clear that at the start the *torque* may be greater than under any other conditions.   This fact is utilised in traction work.

In the case of *the shunt motor* the torque varies with the load in such a way that the speed of rotation remains practically constant within the reasonable working limits of the machine.   As the load is increased one may imagine the speed decreasing, and therefore the back E.M.F. will decrease and the current in the armature will increase, thus increasing the torque and restoring the original speed.   The shunt machine is therefore eminently suited for *constant speed* driving, but is not satisfactory for starting on full load.

The two types then each have their special sphere of usefulness.   The series motor is the machine for traction work—for starting on a heavy load.   But its torque decreases as the speed increases.   The shunt motor is a constant speed machine, its torque so varying with the load that the speed remains constant.

The series motor can be started with a mechanical load on ; but the shunt motor should not be.   It is clear, therefore, that shunt motors are not readily adaptible to tramway work, for example, where large starting torques are essential.

In this work it is common to use two coupled series motors, and to have switch gear, so that they may be connected in series with one another or in parallel. On starting they are connected in series, and when speed has been got up they are connected in parallel. In this way the big rush of current at the start is prevented somewhat, since the total applied E.M.F. is divided through the two motors. And when speed has been got up somewhat and the torque is decreasing they are connected up in parallel, and the current in each is thus increased, and the torque is also increased.

Another plan, adopted in electric motor cars, consists in having one series motors with the armature windings divided on the two commutators ; and with switch gear so that these may be joined up in series on starting, and in parallel when some speed has been attained.

The fact that a shunt motor cannot be used satisfactorily to start with a load on is due to a *demagnetising* tendency of the current in the armature coils. The question of distortion of the field has already been discussed, but it will be seen that the effect of some of the coils on the armature will be to nullify the field due to the field magnets. This is known as the demagnetising effect. With a big load on a shunt motor the starting current in the armature will be very large, and the demagnetising effect will therefore be great. Hence the torque is reduced. But as speed gets up the armature current is decreased and, the demagnetising effect being reduced, the torque actually becomes greater. Of course there will be a limit to this, for the reader will see that a continually reduced current cannot give a continually increasing torque.

In the series motor a larger armature current means a larger field current ; thus the ill effects of the demagnetising coils are not rendered apparent until the field magnets are well on in the saturation stage of magnetisation.

Compound-wound motors are sometimes constructed, but the object of so doing is merely that of regulation, and as such is not very economical.

The point involved is that a series-wound dynamo will run in the reverse direction when run as a motor with the same direction of current. On the other hand, a shunt-wound dynamo will run in the *same* direction as a motor when a given brush is positive in both cases. Of course it must be remembered that the armature current will actually be reversed when a shunt dynamo is used as a motor with a given brush positive. Hence the direction of rotation will be the same.

Clearly then, in a compound-wound motor, the armature will rotate one way if the series field is stronger than the shunt, and the other way if the shunt is stronger than the series, and there will not be any positive gain by compounding.

But with the series coil wound so that it helps the shunt magnetisation the evil effects of demagnetisation due to armature effects may be overcome. This arrangement will amount to a shunt machine with a series winding arranged to increase the *starting* torque.

In order to reverse the direction of rotation of a motor, it is necessary to alter *either* the direction of the armature current, *or* the direction of the current in the field windings, but not both.

In order to vary the speed of a motor, the voltage of supply may be altered, the strength of the field magnetisation may be altered, or the strength of the armature current may be altered. It should be remembered, however, that in general a motor is designed for a definite speed, and that it is most efficient at that speed. Consequently, though an alteration of speed may be essential, it will not be done without some waste of energy, unless the plan of running two direct coupled motors be employed.

There are various methods of altering the quantities mentioned for speed variation, but the discussion of these must be left over.

Hopkinson Test. An interesting method of testing machines is that known as the Hopkinson Test. By means of this the efficiency of a dynamo and a motor can be tested at the same time. Two exactly similar machines are used and one is run as a dynamo and the other as a motor. They are coupled up—either directly or by means of a belt—mechanically ; and leads from the dynamo are connected to the motor.

Now, if friction, and resistance, and hysteresis were non-existent, the current generated by the dynamo would drive the motor, which in turn would mechanically drive the dynamo and perpetual motion would result. But friction, and resist-ance, and hysteresis do exist, and consequently it is necessary to supply additional mechanical energy to the dynamo. The additional energy supplied is a measure of the wasted energy—namely the energy required to overcome the losses due to friction and the rest. This is measured mechanically by a mechanical dynamometer, and the energy given out by the dynamo and absorbed by the motor is measured electrically. Thus the efficiency may be determined. The functions of the two machines may then be reversed.

CHAPTER XV.

ALTERNATING CURRENTS AND TRANSFORMERS.

ALTERNATING currents differ from continuous currents, not only in general effectiveness but also in the fact that they do not follow Ohm's law because of the effects of self-induction.

It has been seen that continuous currents are continuous in the dual sense of their being unidirectional and of unvarying strength with a given resistance. This must be the case with batteries: and continuous current dynamos are so constructed that the same holds good with them. But with an alternator the E.M.F. is continually varying both in direction and magnitude, the latter quantity varying according to the sine curve.

If a non-inductive circuit be supplied with an alternating E.M.F. the alternating current resulting will follow Ohm's law and vary as the E.M.F. varies. That is to say, at any moment the current will be proportional to the applied or *impressed* E.M.F. at that moment. Fig. 135 shews the curves of E.M.F. and current during one complete alternation. The E.M.F. curve is shewn in full line, E.M.F.s being marked off as ordinates and degrees as abscissae. The maximum E.M.F. is assumed to be 100 volts. The current curve is shewn in finer line. The resistance of the non-inductive circuit is assumed to be 2 ohms and the currents are plotted as ordinates.

The current follows the E.M.F. It is said to "keep step" with the E.M.F.; and to be in the *same phase*. When the current strength is varying exactly with the impressed E.M.F. both in magnitude and direction, the two are said to be in the same *phase*.

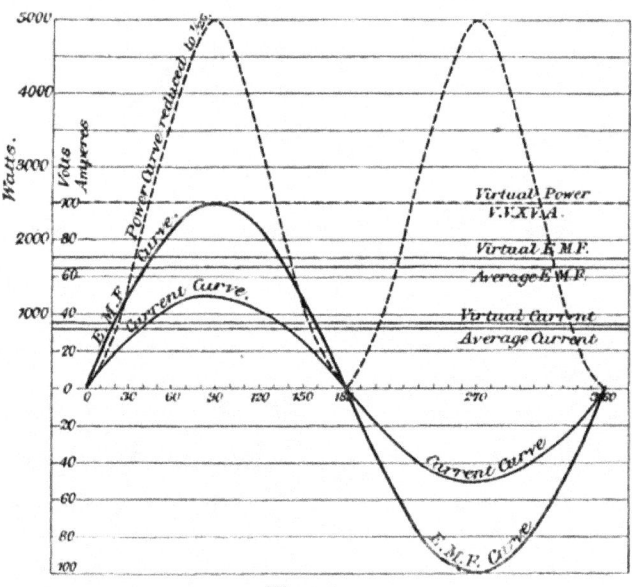

Fig. 135.

Now the average current strength will be 0·637 of the maximum current strength, 0·637 being the average value of the sines of all the angles from 0° to 90°. Thus in the case of the curve illustrated the average current will be 0·637 × 50 = 31·85 amperes; and the average E.M.F. following the same rule will be 0·637 × 100 = 63·7 volts.

However, it is found that an alternating current of 31·85 amperes, that is, a current whose maximum value is 50 amperes will produce a greater heating effect than a continuous current of 31·55 amperes, all other conditions being equal. It would appear therefore that an alternating current is capable of doing more work than an equal continuous current in a given time.

Distinction between square of the average . and average of the squares. It has been shewn that the work done by a current is proportional to the *square* of the current strength, and the difference described above is accounted for by the fact that the square of the average current is not the same thing as the average of the squares of the currents during a quarter of an alternation. In short, the average of the sines of all the angles between 0° and 90° is 0·637. The square of this is 0·405769, or 0·406.

The *average of the squares* of the sines of all angles between 0° and 90° is 0·499849.

Now the quantity 0·406 represents a power factor of a continuous current of 0·637 amperes. The quantity 0·499849 represents the power factor of an alternating current whose average strength is 0·637 amperes. Therefore the alternating current of 0·637 amperes is equivalent to a continuous current of $\sqrt{0·499849} = 0·707$ amperes. Thus it is said that the alternating current whose average strength is 0·637 of its maximum strength is equivalent to a continuous current of 0·707 × that maximum strength. This equivalent current is known as the *virtual amperes.*

On the curve above the average amperes and the virtual amperes are both shewn. The virtual amperes represent the *effectiveness* of the average current, and they are equal to *the square root of the mean square value of the alternating current*; that is to say the square root of the average of the squares of all the sines between 0° and 90° multiplied by the maximum current.

The same applies to alternating E.M.F.s. The average E.M.F. is 0·637 × the maximum E.M.F., but the *virtual E.M.F.*, or *virtual volts*, is 0·707 × the maximum E.M.F.

Average and virtual power. The power in the circuit under discussion will at any instant be the product of the E.M.F. and current. A power curve is shewn on the diagram, the ordinate representing $E \times C$ for each angle. The virtual power is the product of the *virtual amperes* and the *virtual volts.*

Effect of self-induction. In the above discussion it has been assumed that the circuit was non-inductive, in which case the E.M.F. and current followed the sine law strictly and the current was in step with the E.M.F. and followed Ohm's law simply.

But if the circuit is inductive, it follows that there will be self-inductive E.M.F.s, the magnitude and direction of which depend upon the frequency of the current, upon its strength and upon the inductance of the circuit. The general effect of this is that the current ceases to follow the E.M.F. in its rises and falls and *lags* behind. The current at any moment will be that due to the resultant of the impressed E.M.F. and the self-inductive E.M.F., for a given resistance. It will have the same *frequency* as the E.M.F. but it will not be in the same *phase*.

Now the E.M.F. of self-induction will be a maximum when the rate of change of flux is a maximum. That is to say in any inductive circuit the E.M.F. of self-induction is greatest at the instant when the current is changing from zero—at an instant of make or break for example—and gradually decreases as the current grows because the rate of change will decrease.

Therefore, in the case of an alternating current with a continual variation, there will be set up an alternating E.M.F. of induction which will have the same frequency as the current but will not be in the same phase. It will be 90° behind the current since it will be a maximum when the current is a minimum. When the current is a maximum its rate of change must be zero, therefore the induced E.M.F. will be zero. Fig. 136 shews a current curve—following the sine law—and the dotted curve represents the E.M.F. of self-induction. This merely illustrates the variation of the E.M.F. of self-induction with the alternating current.

The E.M.F. which is maintaining this current is called the *active E.M.F.* and this must be the resultant of the *impressed E.M.F.* and the *self-induction E.M.F.* or the counter E.M.F.

Clearly the active E.M.F. must have the same phase as

the current, and thus the self-induction E.M.F. must be 90°
behind this.

But the *impressed* E.M.F. will be out of phase with both
these; and the amount by which it is out must depend upon

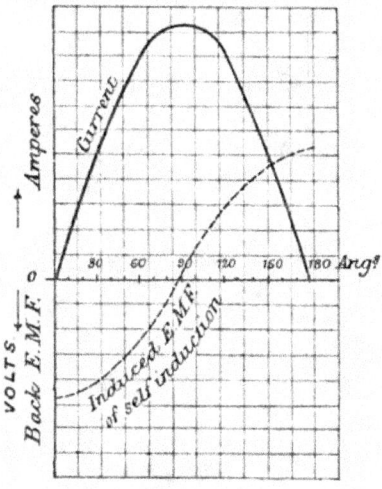

Fig. 136.

the magnitude of the self-induction E.M.F. which will in turn
depend upon the *inductance* of the circuit.

Lag of current behind impressed E.M.F. Figure 137
illustrates all the variations. The current curve was drawn,
and the active E.M.F. curve follows this. The resistance
assumed was 2 ohms; thus the active E.M.F. is just represented by ordinates double the current ordinates. A certain
self-induction was assumed—represented by 50 volts. This
was drawn 90° out of step with the current and active
E.M.F. The *impressed* E.M.F. is obtained by drawing the curve
representing the algebraic sum of the counter E.M.F. and the
active E.M.F.

This, of course, is working backwards, but it serves better
to illustrate the points involved.

The main conclusion of it all is this, that the current *lags behind* the *impressed* E.M.F. which is the E.M.F. applied to the circuit. This lag will depend upon the magnitude of

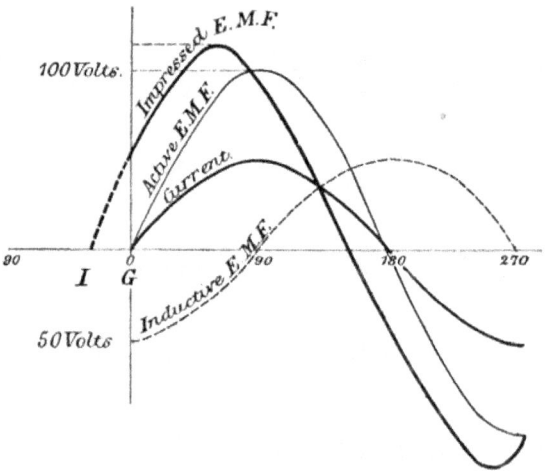

Fig. 137.

the self-induction E.M.F. which in turn will depend upon the induction of the circuit.

The *active* E.M.F. is the *resultant* of the impressed and self-induction E.M.F.s. Its maximum magnitude will always be less than that of the impressed E.M.F.; and by careful inspection of the figure it will be seen that as the lag increases then the maximum magnitude of the active E.M.F. must decrease. It will also be seen that as the E.M.F. of self-induction increases so will the amount of the lag increase, until finally when the self-induced E.M.F. is equal in maximum magnitude to the impressed E.M.F. the lag will be 90° exactly. This will never take place in practice. The *angle of lag* is the angle between the zero value of the impressed E.M.F. and the zero value of the lagging current. In the figure the angle between I and C is the *angle of lag*. This phrase is used solely to denote the lag between the impressed E.M.F. and the current.

The reader should draw curves like those shewn by Fig. 137 taking different values for the current and resistance and the maximum value of the counter E.M.F.

Relationship between impressed, active, and counter E.M.F.s. The general effect then of this self-induction is that the average value of the active E.M.F. is less than that of the impressed E.M.F. Consequently the average current and the virtual current are less than they would be if the circuit was non-inductive and of the same ohmic resistance.

The relationship between the impressed, active and counter E.M.F.s is represented by the relationship between the hypotenuse, base and perpendicular respectively of a right-angled triangle.

In Fig. 138 the base AB represents the active E.M.F. and the perpendicular BC represents the counter E.M.F. which is

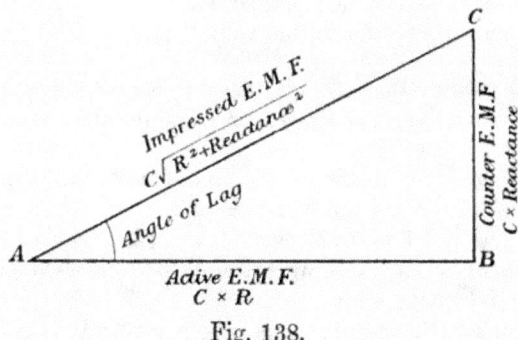

Fig. 138.

90° behind the active and therefore represented as acting at 90° to it. The hypotenuse connecting A and C will represent the impressed E.M.F.

Now the active E.M.F. will be the product of the current and the resistance at any moment. The *virtual* active E.M.F. will be 0·707 × the maximum active E.M.F. It is this which is represented by AB.

The virtual E.M.F. of self-induction will similarly be 0·707 × its maximum value. At any moment it will be

equal to the product of the current and a quantity called
the *reactance*. This quantity depends upon the inductance
of the circuit and it is expressed in terms of resistance, in
ohms. Thus the virtual self-induction E.M.F. = the virtual
current × the reactance.

Now the triangle ABC being right-angled,

$$AB^2 + BC^2 = AC^2,$$

$$\therefore \ AC = \sqrt{AB^2 + BC^2}.$$

$\therefore$ The virtual impressed E.M.F.

$$= \sqrt{(\text{virtual } C \times \text{resis.})^2 + (\text{virtual } C \times \text{reactance})^2}$$

$$= \text{virtual } C \times \sqrt{\text{resis.}^2 + \text{reactance}^2}.$$

Similarly it must follow that

$$\text{Impressed E.M.F.} = \sqrt{\text{active E.M.F.}^2 + \text{inductive E.M.F.}^2},$$

and this will hold good either for instantaneous values, for
average values or for virtual values.

Impedance. The expression $\sqrt{\text{resis.}^2 + \text{reactance}^2}$ is
known as the *impedance*; or sometimes as the *apparent
resistance*.

Hence the square of the impedance is the sum of the
squares of the resistance and the reactance.

Thus it follows that the sides of the triangle of Fig. 138
represent the relationship between the impressed, active and
inductive E.M.F.s, and *also* the relationship between the
impedance, the resistance, and the reactance, of the circuit.

In a circuit conveying an alternating current it therefore
follows that the

$$\text{current} = \frac{\text{impressed E.M.F.}}{\text{impedance}} = \frac{\text{impressed E.M.F.}}{\sqrt{\text{resis.}^2 + \text{reactance}^2}},$$

these being either simultaneous values, or average values.

The angle BAC in the triangle is the *angle of lag* and
it is seen how this angle will increase as the inductive E.M.F.,
BC, increases. It is also seen, however, that it can only be

90° when the inductive E.M.F. is infinite and the impressed E.M.F. will also be infinite.

The angle of lag is such that its tangent is the ratio of the inductive E.M.F. to the active E.M.F.

$$\text{Tan. angle of lag} = \frac{\text{inductive E.M.F.}}{\text{active E.M.F.}}.$$

Reactance. The *reactance* of a circuit depends upon the *frequency* of the alternating current and upon *the coefficient of self-induction* of the circuit, and is directly proportional to these. The *coefficient of self-induction* of a circuit is defined as *the product of the number of lines of force embraced by it when a unit of current is flowing in it and the number of turns in the circuit.*

The practical unit of self-induction is the *Henry,* and the coefficient of self-induction will be expressed in henrys when the current is in amperes and the number of lines of force in C.G.S. units $\div 10^9$. The coefficient of self-induction of a circuit is generally denoted by L.

The *reactance* $\propto$ frequency $\times L$, and

$$\text{reactance in } ohms = 2\pi f L,$$

when L is in henrys and f in alternations per second.

Power in an alternate current system. The rate at which work is being done in a circuit at any instant—namely the power—is measured by the product of E.M.F. and current at any moment. In the case of a continuous current this product would be a constant with a given resistance; but in the case of an alternate current the rate of working will be variable.

Clearly there will be an average rate of working, which will be the product of the average current and the average E.M.F. Also there will be a virtual rate of working, the product of the virtual volts and virtual amperes at a given moment.

The instruments which can be used for measuring alternating currents and E.M.F.s are the hot-wire instruments, the electrostatic instruments and those electro-magnetic instruments which have no iron in their composition, such as the Siemens' dynamometer, the Kelvin balance, and the Siemens' wattmeter. All these instruments will measure virtual amperes, or virtual volts, or virtual watts, because their actions are proportional to squares of currents or to squares of E.M.F.s, or to the product of E.M.F. and current which is in turn proportional to current squared. The hot-wire instrument and the dynamometer are "C^2" instruments ; the electrostatic voltmeter is an "E^2" instrument ; and the wattmeter is, of course, an "$E \times C$" instrument.

In a non-inductive circuit when the current is exactly in phase with the E.M.F. as illustrated by Fig. 135, the power, being the product of the E.M.F. and current at any moment, will follow the curve shewn by the dotted line. Since the E.M.F. and current are in phase the product of these must always be positive—thus the power curve lies entirely on the top side of the base line. This represents positive rate of working.

But when there is inductance in the circuit the current lags behind the impressed E.M.F. as shewn again by Fig. 139.

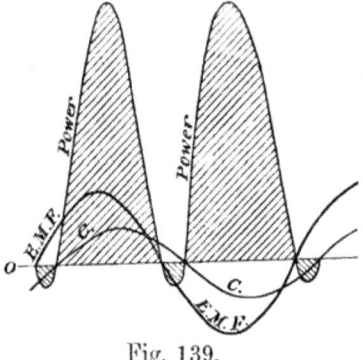

Fig. 139.

Clearly whenever E or C is zero the power will be zero also. It is also seen that at times the current and E.M.F. are in opposite directions, in which case the product must be a negative quantity. The power curve is shewn for this case, the shaded areas being enclosed by it. Above the base line the power is called positive, below it is negative.

Now this negative power is *not* waste. It cannot be applied it is true—but it is non-existent as power. The power expended is represented by the shaded areas above the base line. The shaded areas below the line represent power *given back* by the effects of induction.

Wattless current. As the inductance is increased so will the angle of lag increase with the result that a greater proportion of the power is returned—that is to say appears as "negative" on the power curve. Eventually, if the lag becomes 90° it can be seen that as much power would be returned as was received and that the nett rate of working in that circuit would be *zero*. This case is illustrated by Fig. 140. The current in this case is said to be a *wattless*

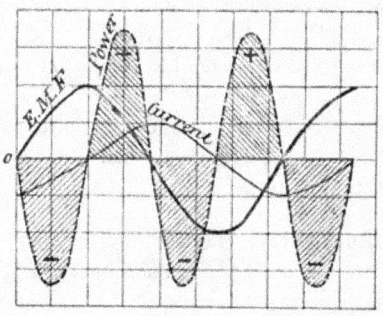

Fig. 140.

current. There is a current flowing, but the power returned to the circuit by the effects of induction are equal to the power received, hence the nett power expended is zero.

The angle of lag can never be exactly 90°, but it can be made to approach it so nearly that the conditions of a wattless current are practically fulfilled. In the use of stationary transformers for alternate current work the primary coil is so designed that when the secondary is "open" there shall be sufficient self-induction to ensure a practically *wattless current*.

Choking coils. Highly inductive coils are frequently introduced into alternating circuits to "choke" down the power. They are called *choking coils* and, though used for the same object as that for which resistance coils are used in continuous current circuits, are nevertheless much more economical in that they do not waste away the energy, but return some by self-induction effects. Of course there will always be the virtual E × virtual C loss in the choker, but the function of the choker is to reduce these values to the necessary minimum. As an illustration of their effectiveness, the following experiment may be performed. A solenoid may be connected to a source of continuous E.M.F. and the resulting current may be measured. The solenoid is next connected to an equal *virtual* alternating E.M.F. and the current is measured again. It will be *less* than before. An iron core, made up of a bundle of soft iron wires, may then be gradually introduced to the interior of the solenoid. It will be found that as the core is put further in the current is reduced more and more. In this way such a solenoid could be used as a choking coil with a variable effectiveness, whilst its actual resistance remains constant.

Effect of capacity. The effect of *capacity* in an alternate current system is to give the current a *lead* in front of the E.M.F. This cannot be discussed in this volume, but the statement may be taken as correct. It may be seen therefore that such a capacity might be introduced into a circuit which would be sufficient to nullify the effects of self-induction—so that the lead tendency of the one would equal the lag tendency of the other—and the current would keep in phase with the impressed E.M.F. as though the circuit was non-inductive.

TRANSFORMERS.

Motor Generator. A *transformer* is a means of transforming an E.M.F. from any given magnitude to any other.
A Ruhmkorff induction coil is a form of transformer for
unidirectional E.M.F.s although it is quite inefficient as a
practical "machine." A more efficient mode of transforming continuous E.M.F.s consists in having two direct-coupled
machines, one of which is wound so that it will run as
motor with the E.M.F. available, and the other wound so
that as a dynamo it will give a higher or lower E.M.F. as may
be desired. This form of transformer is known as a *motor-
generator*. With the ideal transformer the power yielded by
the dynamo should equal the power absorbed by the motor.
But ideal conditions cannot be realised, as in the motor-
generator there is the loss in the motor and also the loss in
the dynamo to be considered. The efficiency of the "set"
will be the ratio of power yielded by the dynamo to that
absorbed by the motor, and this will be less than the efficiency
of either the motor or the dynamo considered separately.

Alternating E.M.F.s may be transformed in a much simpler
manner by means of *stationary* transformers, so called because
there are no moving parts in them.

Object of Transformation. The object of transformation may here be glanced at. A company supplying electric
power to a district has got to face the fact that there must be
a waste of energy in all the feeders, however large these may
be. Further, the cost of large feeders is very considerable.
The power supplied is the product of E and C, so that a given
power may be supplied either as a large current and a small
E.M.F. or as a small current and a large E.M.F. The loss in
the feeders is represented by C^2R, where C is the current in
and R the resistance of the feeders. Clearly then with a
given resistance of feeder it will be more economical to supply
power as a small current at a large E.M.F. But we can go
further than that, for the resistance of the cables may be

increased in inverse ratio, and it will be still more economical, since the heating effect varies directly as the square of the current, but only as the simple resistance.   In this way then the initial cost of laying feeders may be reduced, and at the same time the waste of energy in transmission may also be reduced.   Thus it is that supply companies, especially those feeding a straggling district, generate power at a high pressure and distribute it to transforming stations, where it is transformed down to any required magnitude.

There will of course be some loss in the process of transformation, and in the case of rotary transformers (motor generators or converters), this is very appreciable, and may amount to $20\,°/_°$.   But with alternating E.M.F.s a *stationary* transformer may be used, and the loss of energy in transformation will be very small.   The efficiency of such a transformer may be as high as $98\,°/_°$.

Stationary Transformers.   The stationary transformer consists essentially of two coils, a primary and a secondary, wound on an iron core.   The iron core may be "open" or it may form a "closed" magnetic circuit, the majority being of the latter type.   Faraday's "anchor ring" serves to illustrate

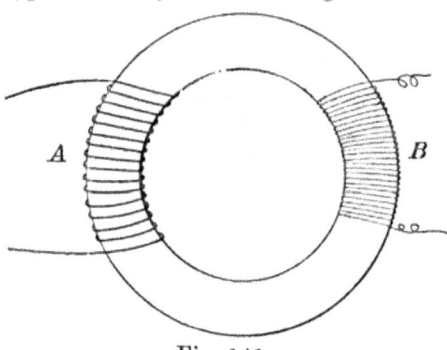

Fig. 141.

the general principle. A and B on Fig. 141 are two coils wound on an iron ring as shewn.   Either coil may be chosen as the primary, but B has more turns of wire than A.

It has already been shewn that the induced E.M.F. in any coil varies as the number of its convolutions for a given rate of change of magnetic flux. It may be assumed that there will be no leakage across the ring, consequently the magnetic flux set up by a current in A will pass through B also. Hence

$$\frac{\text{E.M.F. across ends of } A}{\text{E.M.F. across ends of } B} = \frac{\text{number of } A\text{'s convolutions}}{\text{number of } B\text{'s convolutions}}$$

at each make and break of a current in A.

But if an alternating E.M.F. be applied to A there would be an alternating magnetic flux, and consequently an alternating induced E.M.F. at the ends of B, which would have the same periodicity as that acting in A. Again the ratio of the E.M.F.s in the two coils would be the same as the ratio of their convolutions.

Thus if A be used as the primary coil the E.M.F. in B will be greater than that in A. The transformer would be called a *step-up* transformer. If, however, the coil B be used as the primary the E.M.F. in A will be less than that in B, and the transformer will be used to *step-down*. This is the general use, for the supply companies run their machines at a high voltage and transform down to any required voltage at the consumer's premises.

In the ideal transformer the watts in the secondary would be equal to the watts in the primary; but there must also be some loss in the transformation due to hysteresis, eddy currents, and resistance. Nevertheless the stationary transformer is a most efficient machine, and some large forms may have an efficiency as high as 98—99 %.

The high E.M.F. coil, usually the primary, will only be carrying a relatively small current compared with the secondary low E.M.F. coil. This will necessitate thick wire for the secondary.

The anchor ring of Faraday would not make a satisfactory practical transformer. Firstly the iron core is solid, and as a result eddy currents will be set up when an alternating current flows in the primary coil. Secondly, the coils would be difficult to wind; they could not be machine wound. Thirdly, if the iron ring was covered with winding there would be no means of ventilation.

These are the points which must be looked to in the design of a transformer. The iron must be laminated and insulated in the plane at right angles to the magnetic flux. The iron circuit should be arranged so that separately formed coils could be used; and plenty of cooling surface should be allowed, for the hysteresis loss in the iron cannot be avoided, and the energy wasted appears as heat.

The principle of the *Ferranti* transformer is illustrated by Fig. 142. *P* and *S* are sections of the two coils wound on separate formers or bobbins. They are wound so that the

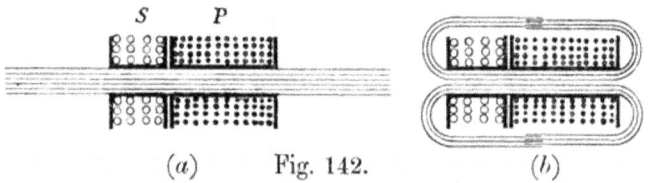

(a) Fig. 142. (b)

ratio of their convolutions gives the necessary ratio of transformation. These formers are slipped on to an iron core composed of a large number of strips of "hoop iron" of Swedish iron. Each strip has been dipped in varnish so that they are insulated from one another. The coils having been slipped along to the centre, the strips of iron are divided and bent over on each side, as shewn by the figure *b*, and the ends are bolted together. In this way eddy currents in the iron are prevented, the coils can be separately wound, and there is a large cooling surface of iron.

The section through a *Mordey* transformer is very similar to the Ferranti so far as the general arrangement is concerned. This is shewn by Fig. 143. The section through the primary windings is shewn by the dots, and through the secondary windings by the small circles. These coils are wound separately on proper formers. The iron part of the transformer is built up of iron stampings. A rectangular plate *abcd* is stamped

24—2

out of soft Swedish iron, and a rectangle *ABCD* is stamped
out of this. The length of the side *AB* of this rectangle is
equal to the width of the original plate—that is to the side
ac; and its width is such that the width of the iron path left
is uniformly the same.

The transformer is built up by slipping an open large rectangle over the formed coils as shewn by the small figure. The
small stamping is slipped *through* the coils; another frame

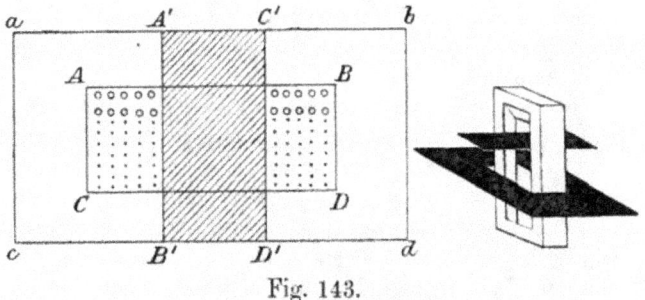

Fig. 143.

rectangle is then slipped over, and another strip put through,
and so on until the whole coil is thus filled up. The stampings
must have been shellaced to prevent eddy currents, and they
are bolted together when the transformer has been completely
built up. There is a large cooling surface of iron.

Many forms are placed in a cast iron box—carefully
insulated from it—and surrounded with oil. This is done
to keep the transformer as cool as possible.

The efficiency of a transformer is measured by the ratio of
the power yielded by the secondary to the power expended in
the primary.

Use of Transformer. Fig. 144 is a diagram which
shews the company's feeders—alternating supply—brought
up through a double pole fuse to the primary coil of a
transformer. The installation to be supplied is connected
up to the secondary of the transformer—also through a
company's main fuse and through the company's meter. The

company's fuse is in a sealed box and the consumer must not break the seal. The meter being on the consumer's side of the transformer only registers the energy yielded by the

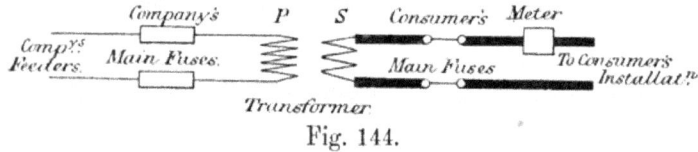

Fig. 144.

secondary coil of the transformer. Consequently when the secondary circuit is open the meter will not give any measurements.

But the primary coil must be connected up the whole time and the E. M. F. of supply must be acting upon it whether the consumer is using the secondary or not. One imagines the company to be losing energy at a considerable rate, but such is not the case. The *self-induction* effects in the primary coil will be large. These effects cause the current to lag behind the E.M. F. and if the inductance be great enough this lag may be made almost 90°. In such a case the amount of energy expended would be equal to that given back by the induction effects, with the result that the nett rate of working would be zero. The current in the primary coil would be what is termed a *wattless current*.

Of course this will never be obtained ; but the current in the primary of a well designed transformer is as nearly *wattless* as possible when the secondary coil is open. This means that there is no appreciable loss of energy in the primary coil when the transformer is not being used : so the company keep their mains connected to it permanently. As soon as the secondary circuit is closed the mutual induction alters the self-induction effects in the primary—reduces them in short. The mutual induction of a closed secondary always tends to lower the self-induction of a primary coil. Hence as the *load* on the secondary is increased, the self-induction of the primary is decreased and consequently the watts absorbed in the primary are increased. In this way the transformer is *self-regulating*.

It only absorbs a number of watts in the primary equal (approximately) to the watts in the secondary—but it does this automatically by the effects of mutual and self-induction. When there is no mutual induction the self-induction causes a *wattless* current : and as the mutual induction increases the current in the primary lags *less* and consequently the watts absorbed become greater whilst the watts returned by self-induction become less.

The stationary transformer thus stands out as a peculiarly simple and efficient machine, requiring no attention and regulating itself without the aid of any mechanism whatever.

ALTERNATING CURRENT MOTORS.

Synchronous Motor. In the ordinary direct current motor the field magnets are magnetised unidirectionally, and the current in the armature coils is reversed every half-turn (in the case of two-polar machines). This reversal is done mechanically by means of a commutator. But it would seem that if an alternating current be supplied to a single-loop armature, then the armature should rotate. However, it can be seen that the alternations must occur at the proper moments, otherwise forward motions would be followed by backward tendencies, and the armature would soon stop. In short, the speed of the armature must be constant, and the number of its revolutions each second must correspond with the number of alternations per second, or with a four-pole machine half this speed, and so on. The speed must "keep step" with the alternations, and this is expressed by saying that the motor speed and the alternations must be *synchronous*.

This is precisely where the trouble of this form of motor comes in. The motor cannot start itself. In the continuous current motor the alternations of the current in the armature are bound to correspond to the speed, because the commutator which causes them is rotating at the same speed as the armature. But with this synchronous motor the alternations of the current supplied are fixed, depending upon the frequency of the alternating E.M.F. used.   The armature is just the same as that of an alternator, and the current is supplied to the brushes making contact on the slip-rings.   There is no commutation—there could be none with the alternating E.M.F. supplied.   Hence with the synchronous motor it must be run up to speed by some external means—by another small direct current motor, for example, which could be arranged with a friction clutch so that it could be uncoupled when the speed had been reached.

When the proper speed is reached the alternating E.M.F. is applied, and the motor will then be driven, provided that it is properly synchronised.   But a small alteration of load and of speed will promptly put it out of step, and it will start pulling up.

In short, the synchronous motor is extremely trouble-some.   Nevertheless, the reader will be able to appreciate its possibilities.   The field magnets must, of course, be separately excited with continuous current; or they may be permanently magnetised.

Induction Motor.   A more general form of *single-phase* alternate current motor is that known as the induction motor. Single-phase alternate current is the ordinary simple alternat-ing current produced by the rotation of a loop in a magnetic field.   A *poly-phase* alternating current is the combination of a number of simple alternating currents of equal frequency, but in different phase at any moment   The induction motor is practically an application of the Arago's disc experiment. If a copper disc be rotated underneath a magnet free to rotate, the magnet will follow up the direction of the disc. If the magnet be rotated the disc will follow up the direction of the magnet.   This is an induction motor, but a wasteful one.   The rotation is produced by "resultant" effects of eddy currents, and the directions of the eddies are naturally not helping the rotation at all points.   The effects may be increased by cutting a lot of slots in the copper disc, leaving a connecting rim at the outside, and of course at the centre.

The disc will be like a spoked wheel with a large number of spokes. In this way the parasitic eddies will be done away with to a large extent, and more efficient results will follow.

In the induction motor proper the field magnets are magnetised with an alternating current. They must be built up of laminated iron therefore. The magnet is fixed, but the effect of the alternating magnetising will be *equivalent to a rotating unidirectional field.*

The magnetisation will follow the alternating current, which follows a sine curve. Hence the field strength across the magnetic axis will vary in exactly the same way as it would do if a uniformly unidirectional field magnet were rotated at a speed corresponding to the frequency of the alternating current. It is thus said that the alternating magnetic field is a *rotary field.*

The armature of the induction motor is *not* supplied with current. It consists, it may be said, of the usual laminated iron core, and copper conductors are laid along it. These are connected together at each end by a copper ring. There are no brushes, and there is no commutator.

The alternating field sets up induced currents in the bars, and as the field amounts to a rotating field the armature rotates, tending all the time to catch up the field, as it were. If the armature came into exact step with the field there would be no alterations in the flux conditions of each conductor. That is the state which the armature looks for ; but, of course, once it has reached it there will be no induced currents set up in it, and it will consequently tend to stop. Thus it would get out of step, and the actions would recommence. But in practice the induction motors are designed to run at a speed midway between this and the speed at which there is the greatest torque.

Laminated Series Motor. Another form of single-phase alternate current motor is the adaptation of the ordinary series-wound direct current machine; the only difference lying in the fact that the field magnets must be built up of laminated iron plates.

With a series motor the direction of rotation will not be altered if the current in *both* field magnet and armature be reversed. This is the principle of the "laminated series motor." However, it is not very satisfactory, except for small machines—something less than 4 horse-power.

The shunt motor cannot be so adapted, because of the enormous self-induction of the field coils.

CHAPTER XVI.

SECONDARY CELLS: ENERGY METERS.

WHEN two plates of dissimilar metals are put into dilute sulphuric acid it is found that there is a difference of potential between them. In the case of the simple cell of Volta the metals used are zinc and copper. If these be pure the dilute acid will have no chemical action upon them, but the difference of potential will exist just the same. It is not known why it exists: it must be regarded merely as one of the phenomena of nature: it is an undoubted fact but it has not yet been explained.

Other metals will produce similar results, though the different pairs of metals will shew different potential differences. The difference of potential appears to vary in some way as the oxidising tendencies of the metals under equal conditions. If these tendencies are exactly equal there will be no difference of potential; and if they are different the difference of potential between them will be greater or less according to the dissimilarity of their oxidising tendencies.

For example the difference of potential between platinum and zinc in dilute sulphuric acid is greater than that between copper and zinc under the same conditions. Between two plates of zinc there will be no potential difference, and the same would apply to any other pair of similar metals.

These facts are applied to the "primary" cells with which the reader is more or less familiar. When the dissimilar metals are connected together by means of a conductor outside the liquid a current flows tending always to equalise the potentials of the metals. This current causes electro-chemical decomposition in the cell, which liberates sufficient energy to maintain the metals at their different potentials. Again it is not known *how* : it is merely known that the total electrical energy obtained from the cell is equal to the chemical energy expended.

When a current is passed through a compound liquid there is a tendency for chemical decomposition. But it can be shewn that a definite E.M.F. is necessary to decompose a given liquid, its magnitude depending upon the liquid used. This corresponds to the *back* E.M.F. of *polarisation*, and this fact is applied in what is known as the *secondary cell*, or the *storage* cell, or the *accumulator*.

In a primary cell of the simple order, like Volta's simple cell, the working is affected by polarisation, which consists broadly in the formation of hydrogen bubbles on the surface of the copper plate. This polarisation increases the internal resistance of the cell, and sets up a back E.M.F. Attempts are made to get rid of this in various ways ; by adding some oxidising agent which will absorb the hydrogen, or by electro-chemical action as in the Daniell cell. It is quite undesirable in the primary cell : it corresponds to the counter E.M.F. of self-induction in alternating circuits.

But Planté saw that it might be used as a means of storing energy ; by changing electrical energy to chemical energy which could be changed back again as desired. It is merely a question of reversibility of action, and the secondary cell may be regarded as a reversible one.

At the present time secondary cells are almost exclusively made of lead plates, coated with a specially prepared layer of lead oxide or sulphate, immersed in sulphuric acid diluted with water in the proportion of one to five.

Two lead plates so formed when placed in the acid will not have different potentials. But if a current be passed through the acid from plate to plate decomposition of the acid will take place, and the liberated constituents will act upon the plates at which they will be liberated. The liberated constituents may be regarded as hydrogen and oxygen. Hydrogen will be freed at the plate where the current *leaves* the liquid, namely, the negative plate, and oxygen at the other. These gases act upon the lead sulphate. The positive plate where the oxygen is liberated will have its sulphate changed into lead superoxide. The lead sulphate on the negative plate will be acted upon by the hydrogen which will reduce it to metallic lead. This metallic lead will be produced in a spongy condition. The whole of the lead sulphate will be thus changed to superoxide of lead and metallic lead if the current be passed for a sufficient length of time.

The result is this : before any current was passed through, the plates were exactly alike, and there was no potential difference. But now the plates are not alike. One is metallic lead whilst the other presents a surface of lead superoxide, and there is, as a result, a difference of potential between them. The lead plate will oxidise more readily than the lead superoxide plate.

Now the process may be exactly reversed. The *charged* cell—it is called *charged* when the plates have been changed as described—has an E.M.F. If therefore its plates be connected together by means of a conductor a current will flow from one to the other outside the cell and in the reverse direction inside.

In all cells it has been found that the current flows from the plate with the smaller oxidising tendency to that with the greater tendency *outside* the cell, and in the reverse direction through the liquid. Thus the superoxidised plate will be the positive and the metallic lead plate will be the negative. But when they are joined together the direction of the current through the acid is the reverse of what it was during the *charging* process, although the terminal of the superoxidised plate is positive to the other in both cases. This may be compared to the direction of the current in the armature of a shunt machine when run as motor and then as dynamo with a given brush positive in both cases.

When the *charged* cell has its plates joined it is said to be *discharging*. The current generated flowing in the reverse direction through the liquid to the charging current decomposes the acid, and the liberated constituents are freed at the opposite plates. Thus the superoxidised plate is reduced, whilst the metallic lead plate is acted upon by the oxidising constituents. Eventually the plates return to their original condition of equality and the cell is discharged.

It is seen then that the actions of charging and discharging are precisely reversible with the ideal secondary cell. And it will be understood that the energy given out on discharge should be equal to that absorbed on charge.

The probable reaction may be represented by

$$\text{Charging} \Big\downarrow \quad \begin{array}{l} PbSO_4 + 2H_2O + PbSO_4 \\ = PbO_2 + 2H_2SO_4 + Pb \end{array} \quad \Big\uparrow \text{Discharging}$$

The E.M.F. of a lead accumulator is finite. It is the E.M.F. between plates of lead and lead superoxide in sulphuric acid. No amount of charging can possibly increase this, and the E.M.F. is independent of the size or internal resistance of the cell.

Other substances would yield different results, but lead gives better results than anything which has yet been tried. The experiment can be tried with copper plates, or zinc plates, or platinum plates, but the E.M.F. produced will be less and the *capacity* will be less. Of course the charging actions will not be the same.

The Capacity of a Secondary Cell. The *capacity* of a secondary cell is measured by the product of current strength and time (expressed as ampere-hours) necessary to completely charge the cell. As soon as the prepared plates have become fully changed to metallic lead and superoxidised plates respectively the cell is charged. After this the decomposed constituents of the acid will be given off as gaseous products, since they will no longer be absorbed by the plates. When

this action commences the cell is said to be "gassing," and
the energy expended in decomposing the acid is being wasted.
The capacity of the cell will therefore depend upon the
effective area of its plates; that is upon the amount of
energy which will be necessary to change the prepared lead
plates to the forms described. A cell of 30 ampere-hours
capacity requires a current of 30 amperes for one hour, or
10 amperes for three hours, etc., to charge it. Conversely in
discharging it will yield a current of 30 amperes for an
hour, etc.

The *charging* and *discharging rate* is expressed by the
maximum current strength it is desirable to pass in charging
or discharging. It would not be desirable, merely for practical
reasons, to charge the above 30 ampere-hour cell with a
current of 30 amperes for an hour, or to discharge it at that
rate. It has been found that large currents tend to dis-
integrate the plates, after which there will be a tendency for
them to "buckle." The maximum charging rate is usually
about 0·025 amperes per square inch of plate. The maximum
rate of discharging is usually the same as that of charging;
but in certain types of cells this may be increased. The
manufacturers generally supply the exact rate with a given
cell.

Conditions of Charging. The E.M.F. of the charging
current must, of course, be greater than the back E.M.F. of
the cell; and in this connexion it should be remembered that
the E.M.F. of the cell rises rapidly to 1·9 volts and then more
or less gradually, according to the capacity, up to about 2·5
volts. Thus the charging E.M.F. must be at least 2·5 volts
per cell in the case of a series battery. In any cell during
charging the current strength

$$C = \frac{E - e}{b},$$

where E is the charging E.M.F., e the E.M.F. of the cell, acting
backwards of course, and b the internal resistance. A cell
should never be allowed to stand for any length of time in a

discharged state. Even charged cells not being used should be given a slight charge every week. There appears to be some kind of local action which tends to discharge a cell.

Discharging. In discharging, the E.M.F. of each cell falls rapidly to about 2 volts, and then slowly, according to capacity, to 1·9 volts. The E.M.F. should not be allowed to fall below this, and the cell should be recharged.

Fig. 145 *a* and *b* are charging and discharging curves, illustrating the variations of E.M.F. with the ampere-hours.

If the cells are discharged at too great a rate a white scaly film forms over the plates. This is probably the sulphate Pb_2SO_5, and a cell in this condition is said to be "sulphated."

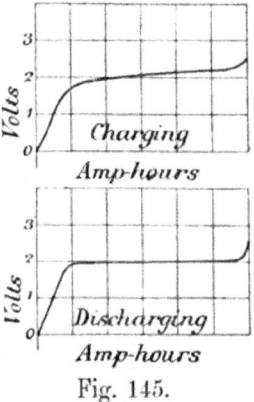

Fig. 145.

This is an ambiguous term of course, but it is used only in reference to this particular formation, which renders the plates inactive.

The plates of an accumulator are so constructed that they are capable of holding the paste thoroughly and at the same time present as large a surface as is possible. The chief troubles of secondary cells arise from tendency of the paste to fall off the plates—known as disintegration. This may be brought about by too great a rate of discharge or of charge.

Disintegration is frequently followed by "buckling" of the plates and consequent internal short-circuiting.

Since the actions on the two sets of plates are different, they are usually made of different forms. The negative plate does not have the same tendencies towards disintegation as the positive plate. One form of positive plate is shewn by Fig. 146*a*. It consists of a lead plate with a number of projecting flat ridges inclined upwards to the plate. The paste is packed in these. The negative plate becomes reduced to

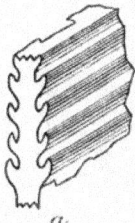 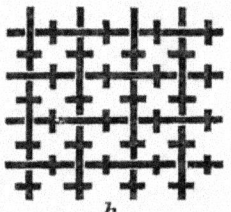

Fig. 146.

lead in a spongy condition, and the actions are productive of greater changes of volume of the formed plate. This plate is often made in the form of an open grid, such as that shewn by 146*b*. The projecting tongues serve to hold the paste in position. These are plates used in the E.P.S. cell.

In a single cell, it is usual to have a number of plates, with one more positive plate than negative. These are arranged alternately, parallel to one another, there being a positive plate at each end of the series. All the negative plates are connected up to a lead cross bar which has a projecting lug for coupling purposes. The positive plates are similarly connected to another bar and lug.

The capacity of the cell is increased by having a greater number of plates. But, of course, the E.M.F. is not altered. The internal resistance is decreased by increasing the number of plates, and this is a point of extreme importance in the practical running of accumulators.

The plates must be kept apart, and this is generally done

by means of ebonite forks which are slipped over the positive plate, one prong being each side of the plate.

The containing vessel may be glass, or ebonite, or celluloid, or lead lined teak. In the case of lead lined teak cases the plates must not come into contact with the lead lining, neither must the connecting bars. Ebonite is again used to prevent this.

Accumulators to be used for traction work—in automobiles, tramways, or light railways—must have protecting lids so that there shall be no spilling of the acid.

The great drawback to accumulators for traction lies in the fact that they are so very heavy for a limited capacity. Many problems of locomotion will be solved on the day when an accumulator is designed which shall have a large capacity and a very small weight.

The acid used in accumulators should have a relative density of 1·190 when the cell is uncharged primarily. On complete charge it will be found that this value has risen to 1·220. This is due to the fact that in decomposition some of the water has been removed, as such. On discharging this will be returned and the relative density will drop. Special *hydrometers* are manufactured for use in secondary cells. The relative density of the acid enables one to form an estimate of the condition of a cell with respect to its charge.

The acid should never be lower than half an inch above the tops of the plates. Due to evaporation of the water and the "spraying" during action the level will decrease. The density of the acid will be greater then than it ought to be. In bringing up the level again it will generally be necessary to add *distilled* water only. But, of course, finally the density of acid should be adjusted to its proper value. To reduce the losses due to spraying it is usual to have a curved glass plate lying over the liquid, the convex surface being towards the acid.

Spraying and creeping of the acid are a source of trouble at the connecting terminals unless the latter are coated with paraffin wax or something equivalent.

Efficiency of secondary cells. The efficiency of a secondary cell is a variable quantity. It varies as the rate of discharge, and it may be stated that the efficiency is greatest when the rate of discharge is least. The maximum efficiency is probably not greater than 75 °/₀ under ordinary conditions of working; but with special conditions an efficiency of 90 °/₀ may be obtained. By "efficiency of a secondary cell" one means the ratio of the total energy given out by it in discharging to that put into it in charging.

Since they can be made with very low internal resistances accumulators may be used efficiently for lighting purposes. They are usually used as a "stand-by" in conjunction with an engine and dynamo. When the lighting load is small it is run off the cells, but when it is heavy then the dynamo is run up. Usually, too, when the dynamo is running, arrangements are made to charge the cells at the same time.

They serve excellently too as a means of regulating the E.M.F. of a generator driven by a gas-engine. Such an engine causes a variable E.M.F. to be generated and this would be shewn by variation in the candle-power of the illuminating lamps. The secondary battery is joined up in parallel with the dynamo—a sufficient number of cells being used in series with each other to yield the necessary E.M.F.

For use in this work there are many automatic "cut-outs" designed which automatically switch the dynamo out of the circuit if it should cease to work properly. These cut-outs work on electromagnetic principles.

Energy meters. An ammeter and a voltmeter can give no indication of the amount of electrical energy which is expended in any circuit during some period of time. It is true that the product of volts and amperes indicate the rate at which work is being done at the instant at which the readings are taken; and further if these remained constant over a given period of time, the product of the watts and the time in seconds would express the total work done in joules.

It has already been explained that electrical energy is

bought and sold at so much per Board of Trade unit, that unit being a kilowatt-hour. This represents a definite quantity of work. One watt-second is a joule, and thus a *kilowatt-hour* is 3,600,000 *joules*. This is the unit called the Board of Trade unit.

The *meters* used for the measurement of electrical energy are broadly of two classes. The one class are known as *ampere-hour meters*, whilst the other class are *watt-hour meters* or *unit meters*.

The ampere-hour meter records the product of amperes and hours. Thus with a current of 20 amperes for 6 hours a reading of 120 ampere-hours would be recorded. The same record would shew with a current of 5 amperes for 24 hours ; or 60 amperes for 2 hours ; or for a current of 10 amperes for 2 hours and then 20 amperes for 5 hours, and so on.

Clearly an ampere-hour meter must be a recording meter of some kind. Furthermore, ampere-hours do not give a measure of work. They are merely a measure of the total *quantity* of electricity. One ampere-hour is equivalent to 3600 coulombs of electricity.

The work done by this quantity will be found by its product with the E.M.F. which was urging it.

It is clear therefore that an indication of one ampere-hour has no significance so far as work is concerned unless the urging E.M.F. be known.

Thus an ampere-hour meter registers as having passed through it a quantity of, say, 20 ampere-hours.

If the urging E.M.F. was 100 volts—unvarying—the work done would be $20 \times 100 = 2000$ volt-ampere-hours, that is to say would be 2000 watt-hours.

But a Board of Trade unit is a kilowatt-hour, and therefore the work done in Board of Trade units would be

$$\frac{2000}{1000} = 2 \text{ units.}$$

If the urging E.M.F. had been 230 volts, then with the *same* ampere-hour reading the work done would have been

$$\frac{20 \times 230}{1000} = 4\cdot6 \text{ Board of Trade units.}$$

Ampere-hour meters.   Ampere-hour meters are graduated to record either ampere-hours or Board of Trade units.
In the former case the work done will be found by multiplying
the ampere-hours by the urging E. M. F. and dividing the product
by 1000.

In the latter case the meter is designed to be used for
a definite E.M.F., and the dials are graduated to read Board
of Trade units directly.   If the E. M. F. used is constant and is
of the magnitude for which the meter is made these readings
will be correct.   But if some other E. M. F. be used then the
readings will be too high or too low as the E. M. F. is greater or
less than that for which the meter is to be used.

A simple form of continuous current ampere-hour meter is
one which consists of a light armature—drum wound and
commutated—which can rotate between the poles of a permanent horseshoe magnet.   The spindle is arranged vertically
and the armature is wound with fine high resistance coils.
The brushes are connected to the ends of a low resistance,
which is connected in series with the circuit in which the
ampere-hours are to be measured.

The meter is thus a small motor of high resistance which
is shunted across a fixed low resistance after the manner of
the movable coil ammeters (see page 54).   The speed of the
motor is directly proportional to the current in the armature,
since the field magnets are permanent, and the current in the
armature must be proportional to the difference of potential
across the ends of the shunt, which in turn is proportional to
the main current in the circuit.   Thus the speed of the motor
is proportional to the main current.

It should here be noted that to attain this *either* there
must be no iron core to the armature *or* the winding must be
such that the iron is not magnetised to saturation at any time
but is always on the straight part of its magnetisation curve.

The speed being proportional to the main current it follows
that the total number of revolutions made in a given time will
be proportional to the ampere-hours transmitted during that
time.   To record this the spindle is geared on to a train of

wheels with a counting mechanism very much like that of an ordinary gas meter, and the dials are marked off to read ampere-hours in units, tens, hundreds and so on, depending upon the size of the meter.

When it is graduated to register Board of Trade units the dials are marked differently. A definite E.M.F. is assumed, and the ampere-hours × this E.M.F. and ÷ 1000 will give a corresponding reading in Board of Trade units.

But if this be used for any other E.M.F. the readings will be false. A little consideration will shew however that the true reading of Board of Trade units will be obtained from the following.

Actual reading of B.O.T. units

$$\times \frac{\text{E. M. F. used}}{\text{designed E. M. F.}} = \text{true reading.}$$

If a supply company installed an ampere-hour meter graduated to read units on a 100 volt supply and they supplied 230 volts it can be seen that the consumer would only be paying for $\frac{1}{2 \cdot 3}$ of the total units which he received.

Ferranti meter. Another form of ampere-hour meter is that known as the Ferranti meter. The principle of this is illustrated by Fig. 147. Some mercury is placed in a shallow iron vessel of circular form as shewn and a current is passed through it from the centre to the circumference, an insulated conductor leading the current up through the vessel to the mercury at the centre. The current will radiate in all directions from the centre. Above this mercury trough a solenoid or an electromagnet is arranged so that it projects lines of force at right angles to the surface of the mercury. It is assumed that they are projected downwards. Now the current tends to move *across* the lines of force, and the result is that the whole mass of the mercury will rotate—in this case in a left-handed direction.

The solenoid is connected in series with the mercury trough so that the forces of rotation will be proportional to the *current squared*, since they are proportional to the current in the mercury and the current in the solenoid.

A disc floats on the mercury and this disc has a vertical spindle which is geared up to a counting and recording

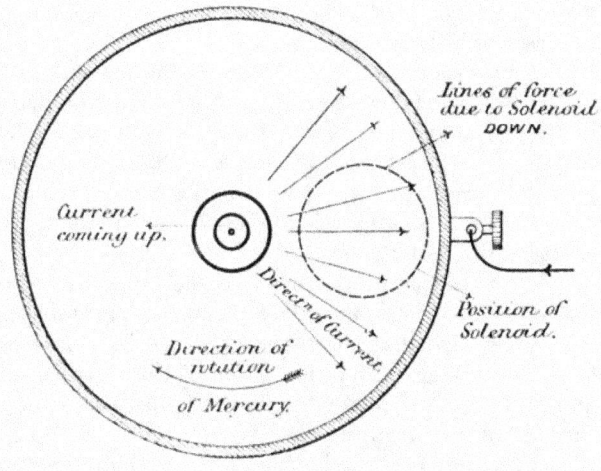

Fig. 147.

mechanism, graduated in ampere-hours or Board of Trade units as occasion demands.

Thomson kilowatt-hour meter.   One form of kilowatt-hour meter which records the true kilowatt-hours even with a varying pressure is that designed by Prof. Elihu Thomson and generally known as the Thomson meter.   It can be used for either continuous or alternating power supplies.

It is practically a small motor without any iron.   The armature is high resistance drum-wound on a non-magnetic core, commutated with a silver commutator.   It turns about a vertical axis, the spindle running in jewels at top and bottom.   In series with the armature is a high resistance, and these are connected across the mains, so that the small current which will pass through the armature will be proportional to the E.M.F. of supply.

The "field magnets" of this meter motor consist of two low resistance thick wire coils, placed with their planes vertical and parallel to each other and on opposite sides of the armature. They are connected in series so that the current will be going in the same direction in each at the same moment. There is no iron core. The "field magnet" coils are connected in series with the main circuit.

Now the motor will rotate when the connexions are made. The torque will be proportional to the product of the current in the armature and that in the field coils. Therefore the torque will be proportional to $E \times C$. But if there are practically no forces to overcome the motor would simply "race" away with an acceleration and it could hardly be used as an index of work done.

A thin copper disc is fixed to the armature spindle, so that its plane is horizontal and a permanent horse-shoe magnet is placed so that the plate rotates between its poles when the meter is working. Now the effects of the eddy currents set up in the disc will retard the motor. The strength of these currents varies directly as the speed since the field is constant. Hence as the motor tends to go quicker the retardation due to eddy currents in the disc becomes greater.

Therefore it follows that the number of revolutions in a given time will be a measure of the *work done* in that time. That is to say the speed gives a measure of the rate of working, and the total number of revolutions in any time gives a measure of the watt-hours of work.

The spindle is geared on to a counting and recording mechanism. This records Board of Trade units or in some cases it merely records the number of revolutions and a "constant" is given, the product of which with the number of revolutions is the Board of Trade units.

Fig. 148 illustrates the method of connecting the watt-hour meter up in a system.

The effects of friction, which might introduce a considerable error in the long run, are compensated for by an additional winding on the field coils of some turns in series with the

armature. The object of this is to get the meter to start
with the smallest current in the main circuit.

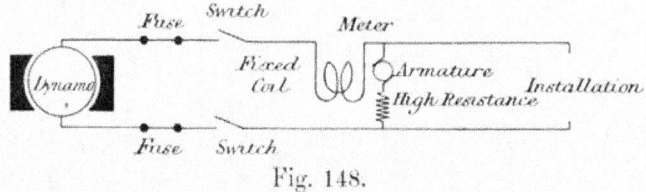

Fig. 148.

Graphically recording meters. A large number of
meters are now made which record graphically the ampere-
hours. One may get the general idea by imagining a pen
attached to the points of an ordinary ammeter and so arranged
that it rests on a cylinder which can be rotated by means
of a clock. On the cylinder some paper is rolled—this
having been ruled, say, with lines which are "1 hour apart"
—that is to say the drum will turn so that the pen passes
from one line to the next in one hour. In this way the pen
will move across the drum as the current varies and it is
being moved around the drum uniformly with the time.
The graph obtained will shew the amperes at every moment,
and the mean current × time in hours will give the ampere-
hours.

Recording voltmeters are also made on similar lines and
the combined records of one of these with a recording ammeter
will furnish the watt-hours and the Board of Trade units.

APPENDIX.

TABLE OF NATURAL SINES AND TANGENTS.

Angle	Sine	Tangent	Angle	Sine	Tangent
0	0	0	31	+ ·515	+ ·601
1	+ ·018	+ ·018	32	+ ·530	+ ·625
2	+ ·035	+ ·035	33	+ ·545	+ ·649
3	+ ·052	+ ·052	34	+ ·559	+ ·674
4	+ ·070	+ ·070	35	+ ·574	+ ·700
5	+ ·087	+ ·088	36	+ ·588	+ ·726
6	+ ·104	+ ·105	37	+ ·602	+ ·754
7	+ ·122	+ ·123	38	+ ·616	+ ·781
8	+ ·139	+ ·140	39	+ ·629	+ ·810
9	+ ·156	+ ·158	40	+ ·643	+ ·839
10	+ ·174	+ ·176	41	+ ·656	+ ·869
11	+ ·191	+ ·194	42	+ ·669	+ ·900
12	+ ·208	+ ·213	43	+ ·682	+ ·932
13	+ ·225	+ ·231	44	+ ·695	+ ·966
14	+ ·242	+ ·249	45	+ ·707	+ 1·000
15	+ ·259	+ ·268	46	+ ·719	+ 1·036
16	+ ·276	+ ·287	47	+ ·731	+ 1·072
17	+ ·292	+ ·306	48	+ ·743	+ 1·111
18	+ ·309	+ ·325	49	+ ·755	+ 1·150
19	+ ·326	+ ·344	50	+ ·766	+ 1·192
20	+ ·342	+ ·364	51	+ ·777	+ 1·235
21	+ ·358	+ ·384	52	+ ·788	+ 1·280
22	+ ·375	+ ·404	53	+ ·799	+ 1·327
23	+ ·391	+ ·424	54	+ ·809	+ 1·376
24	+ ·407	+ ·445	55	+ ·819	+ 1·428
25	+ ·423	+ ·466	56	+ ·829	+ 1·483
26	+ ·438	+ ·488	57	+ ·839	+ 1·540
27	+ ·454	+ ·510	58	+ ·848	+ 1·600
28	+ ·470	+ ·532	59	+ ·857	+ 1·664
29	+ ·485	+ ·554	60	+ ·866	+ 1·732
30	+ ·500	+ ·577	61	+ ·875	+ 1·804

Angle	Sine	Tangent	Angle	Sine	Tangent
62	+ ·883	+ 1·881	105	+ ·966	− 3·732
63	+ ·891	+ 1·963	106	+ ·961	− 3·487
64	+ ·899	+ 2·050	107	+ ·956	− 3·271
65	+ ·906	+ 2·144	108	+ ·951	− 3·078
66	+ ·914	+ 2·246	109	+ ·946	− 2·904
67	+ ·920	+ 2·356	110	+ ·940	− 2·748
68	+ ·927	+ 2·475	111	+ ·934	− 2·605
69	+ ·934	+ 2·605	112	+ ·927	− 2·475
70	+ ·940	+ 2·748	113	+ ·920	− 2·356
71	+ ·946	+ 2·904	114	+ ·914	− 2·246
72	+ ·951	+ 3·078	115	+ ·906	− 2·144
73	+ ·956	+ 3·271	116	+ ·899	− 2·050
74	+ ·961	+ 3·487	117	+ ·891	− 1·963
75	+ ·966	+ 3·732	118	+ ·883	− 1·881
76	+ ·970	+ 4·011	119	+ ·875	− 1·804
77	+ ·974	+ 4·332	120	+ ·866	− 1·732
78	+ ·978	+ 4·705	121	+ ·857	− 1·664
79	+ ·982	+ 5·145	122	+ ·848	− 1·600
80	+ ·985	+ 5·671	123	+ ·839	− 1·540
81	+ ·988	+ 6·314	124	+ ·829	− 1·483
82	+ ·990	+ 7·115	125	+ ·819	− 1·428
83	+ ·992	+ 8·144	126	+ ·809	− 1·376
84	+ ·994	+ 9.514	127	+ ·799	− 1·327
85	+ ·996	+11·430	128	+ ·788	− 1·280
86	+ ·997(6)	+14·301	129	+ ·777	− 1·235
87	+ ·998(6)	+19·081	130	+ ·766	− 1·192
88	+ ·999(4)	+28·636	131	+ ·755	− 1·150
89	+ ·999(8)	+57·290	132	+ ·743	− 1·111
90	+1·000	+ ∞	133	+ ·731	− 1·072
91	+ ·999(8)	−57·290	134	+ ·719	− 1·036
92	+ ·999(4)	−28·636	135	+ ·707	− 1·000
93	+ ·998(6)	−19·081	136	+ ·695	− ·966
94	+ ·997(6)	−14·301	137	+ ·682	− ·932
95	+ ·996	−11·430	138	+ ·669	− ·900
96	+ ·994	− 9·514	139	+ ·656	− ·869
97	+ ·992	− 8·144	140	+ ·643	− ·839
98	+ ·990	− 7·115	141	+ ·629	− ·810
99	+ ·988	− 6·314	142	+ ·616	− ·781
100	+ ·985	− 5·671	143	+ ·602	− ·754
101	+ ·982	− 5·145	144	+ ·588	− ·726
102	+ ·978	− 4·705	145	+ ·574	− ·700
103	+ ·974	− 4·332	146	+ ·559	− ·674
104	+ ·970	− 4·011	147	+ ·545	− ·649

Angle	Sine	Tangent	Angle	Sine	Tangent
148	+ ·530	− ·625	191	− ·191	+ ·194
149	+ ·515	− ·601	192	− ·208	+ ·213
150	+ ·500	− ·577	193	− ·225	+ ·231
151	+ ·485	− ·554	194	− ·242	+ ·249
152	+ ·470	− ·532	195	− ·259	+ ·268
153	+ ·454	− ·510	196	− ·276	+ ·287
154	+ ·438	− ·488	197	− ·292	+ ·306
155	+ ·423	− ·466	198	− ·309	+ ·325
156	+ ·407	− ·445	199	− ·326	+ ·344
157	+ ·391	− ·424	200	− ·342	+ ·364
158	+ ·375	− ·404	201	− ·358	+ ·384
159	+ ·358	− ·384	202	− ·375	+ ·404
160	+ ·342	− ·364	203	− ·391	+ ·424
161	+ ·326	− ·344	204	− ·407	+ ·445
162	+ ·309	− ·325	205	− ·423	+ ·466
163	+ ·292	− ·306	206	− ·438	+ ·488
164	+ ·276	− ·287	207	− ·454	+ ·510
165	+ ·259	− ·268	208	− ·470	+ ·532
166	+ ·242	− ·249	209	− ·485	+ ·554
167	+ ·225	− ·231	210	− ·500	+ ·577
168	+ ·208	− ·213	211	− ·515	+ ·601
169	+ ·191	− ·194	212	− ·530	+ ·625
170	+ ·174	− ·176	213	− ·545	+ ·649
171	+ ·156	− ·158	214	− ·559	+ ·674
172	+ ·139	− ·140	215	− ·574	+ ·700
173	+ ·122	− ·123	216	− ·588	+ ·726
174	+ ·104	− ·105	217	− ·602	+ ·754
175	+ ·087	− ·088	218	− ·616	+ ·781
176	+ ·070	− ·070	219	− ·629	+ ·810
177	+ ·052	− ·052	220	− ·643	+ ·839
178	+ ·035	− ·035	221	− ·656	+ ·869
179	+ ·018	− ·018	222	− ·669	+ ·900
180	0	0	223	− ·682	+ ·932
181	− ·018	+ ·018	224	− ·695	+ ·966
182	− ·035	+ ·035	225	− ·707	+1·000
183	− ·052	+ ·052	226	− ·719	+1·036
184	− ·070	+ ·070	227	− ·731	+1·072
185	− 087	+ ·087	228	− ·743	+1·111
186	− 104	+ ·105	229	− ·755	+1·150
187	− ·122	+ ·123	230	− ·766	+1·192
188	− ·139	+ ·140	231	− ·777	+1·235
189	− ·156	+ ·158	232	− ·788	+1·280
190	− ·174	+ ·176	233	− ·799	+1·327

Angle	Sine	Tangent	Angle	Sine	Tangent
234	− ·809	+ 1·376	277	− ·992	− 8·144
235	− ·819	+ 1·428	278	− ·990	− 7·115
236	− ·829	+ 1·483	279	− ·988	− 6·314
237	− ·839	+ 1·540	280	− ·985	− 5·671
238	− ·848	+ 1·600	281	− ·982	− 5·145
239	− ·857	+ 1·664	282	− ·978	− 4·705
240	− ·866	+ 1·732	283	− ·974	− 4·332
241	− ·875	+ 1·804	284	− ·970	− 4·011
242	− ·883	+ 1·881	285	− ·966	− 3·732
243	− ·891	+ 1·963	286	− ·961	− 3·487
244	− ·899	+ 2·050	287	− ·956	− 3·271
245	− ·906	+ 2·144	288	− ·951	− 3·078
246	− ·914	+ 2·246	289	− ·946	− 2·904
247	− ·920	+ 2·356	290	− ·940	− 2·748
248	− ·924	+ 2·475	291	− ·934	− 2·605
249	− ·934	+ 2·605	292	− ·927	− 2·475
250	− ·940	+ 2·748	293	− ·920	− 2·356
251	− ·946	+ 2·904	294	− ·914	− 2·246
252	− ·951	+ 3·078	295	− ·906	− 2·144
253	− ·956	+ 3·271	296	− ·899	− 2·050
254	− ·961	+ 3·487	297	− ·891	− 1·963
255	− ·966	+ 3·732	298	− ·883	− 1·881
256	− ·970	+ 4·011	299	− ·875	− 1·804
257	− ·974	+ 4·332	300	− ·866	− 1·732
258	− ·978	+ 4·705	301	− ·857	− 1·664
259	− ·982	+ 5·145	302	− ·848	− 1·600
260	− ·985	+ 5·671	303	− ·839	− 1·540
261	− ·988	+ 6·314	304	− ·829	− 1·483
262	− ·990	+ 7·115	305	− ·819	− 1·428
263	− ·992	+ 8·144	306	− ·809	− 1·376
264	− ·994	+ 9·514	307	− ·799	− 1·327
265	− ·996	+11·430	308	− ·788	− 1·280
266	− ·997(6)	+14·301	309	− ·777	− 1·235
267	− ·998(6)	+19·081	310	− ·766	− 1·192
268	− ·999(4)	+28·636	311	− ·755	− 1·150
269	− ·999(8)	+57·290	312	− ·743	− 1·111
270	−1·000	+ ∞	313	− ·731	− 1·072
271	− ·999(8)	−57·290	314	− ·719	− 1·036
272	− ·999(4)	−28·636	315	− ·707	− 1·000
273	− ·998(6)	−19·081	316	− ·695	− ·966
274	− ·997(6)	−14·301	317	− ·682	− ·932
275	− ·996	−11·430	318	− ·669	− ·900
276	− ·994	− 9·514	319	− ·656	− ·869

Angle	Sine	Tangent	Angle	Sine	Tangent
320	− ·643	− ·839	341	− ·326	− ·344
321	− ·629	− ·810	342	− ·309	− ·325
322	− ·616	− ·781	343	− ·292	− ·306
323	− ·602	− ·754	344	− ·276	− ·287
324	− ·588	− ·726	345	− ·259	− ·268
325	− ·574	− ·700	346	− ·242	− ·249
326	− ·559	− ·674	347	− ·225	− ·231
327	− ·545	− ·649	348	− ·208	− ·213
328	− ·530	− ·625	349	− ·191	− ·194
329	− ·515	− ·601	350	− ·174	− ·176
330	− ·500	− ·577	351	− ·156	− ·158
331	− ·485	− ·554	352	− ·139	− ·140
332	− ·470	− ·532	353	− ·122	− ·123
333	− ·454	− ·510	354	− ·104	− ·105
334	− ·438	− ·488	355	− ·087	− ·088
335	− ·423	− ·466	356	− ·070	− ·070
336	− ·407	− ·445	357	− ·052	− ·052
337	− ·391	− ·424	358	− ·035	− ·035
338	− ·375	− ·404	359	− ·018	− ·018
339	− ·358	− ·384	360	− 0	− 0
340	− ·342	− ·364			

USEFUL DATA.

$$\pi = 3\cdot14(159).$$
$$\sqrt{2} = 1\cdot414.$$
$$\sqrt{3} = 1\cdot732.$$

1 metre	= 39·37 inches.
1 cm.	= ·3937 ,,
1 inch	= 2·54 cms.
1 sq. inch	= 6·45 sq. cms.
1 cu. inch	= 16·38 cu. cms.
1 lb.	= 453·6 grams.
1 kilogram	= 2·205 lbs.
1 gram	= ·0022 lbs.
1000 c.c.	= 1 litre.
1 litre	= 1·76 English pints.

1 c.c.	of water weighs	1 gram.	
1 litre	,, ,,	7 kilograms.	
1 gallon	,, ,,	10 lbs.	
1 cu. foot	,, ,,	62·3 lbs. = 1000 ozs. (appr.).	

1 normal atmosphere = 14·7 lbs. per sq. inch.
Normal barometric height for mercury = 30 ins. = 76 cms.
,,   ,,   ,,   water = 34 feet.

$$\text{Area of a triangle} = \frac{\text{base} \times \text{altitude}}{2}.$$

Area of a circle $= (\text{radius})^2 \times \pi.$

Area of the surface of a sphere $= (\text{radius})^2 \times 4\pi.$

Volume of a cylinder of height h and radius of base $r = \pi r^2 \cdot h.$

Volume of a sphere $= \frac{4}{3}\pi \cdot (\text{radius})^3.$

1 horse-power = 746 watts.

1 B.O.T. unit = 1000 watt-hours = 1 kilowatt-hour.

Joule's mechanical equiv. of Heat

$= 4\cdot2 \times 10^7$ ergs per gram Centigrade unit

$= 778$ ft.-lbs. per pound Fahrenheit unit.

Conversion of Temperature Scales.

(Degrees $F - 32$) $\times \frac{5}{9}$ = Degrees Centigrade.

(Degrees $C \times \frac{9}{5}$) $+ 32$ = Degrees Fahrenheit.

QUESTIONS.

1. What is the distinction between the E.M.F. of a cell and the difference of potential between its terminals when it is generating a current? Upon what do they depend?

2. Two cells, each of E.M.F. 1·1 volts are connected separately to resistances of 4 and 5 ohms. If the cells have equal internal resistances of 1·5 ohms what will be the potential difference between their terminals?

3. Two cells A and B have E.M.F.s of 1·46 and 1·98 volts respectively. A is connected to a resistance of 7 ohms and B to one of 10 ohms. The current strength is found to be 0·146 amp. in both cases. What is the internal resistance of each cell, and what is the difference of potential between their terminals whilst the current is flowing?

4. If, in the above question, the internal resistance of each cell was 1·5 ohms what would be the current strength in each circuit, and which would be the more efficient: regarding the volts used in urging the current through each cell as waste?

5. When a cell is connected up to a resistance and a voltmeter is connected to the ends of that resistance, what do its readings indicate? How do you account for the fact that as the resistance is increased the voltmeter readings also increase? When will the limit be reached?

6. Explain, as fully as you can, how a volt can be defined in terms of *work*.

7. Describe any form of ammeter, pointing out any special advantages or disadvantages which it may have.

8. What are the essential conditions of a good ammeter, and why?

9. Enumerate the disadvantages of "iron-needle ammeters."

10. In what respects do galvanometers differ from ammeters?

11. A Siemens' dynamometer is being standardised by means of a silver voltameter. It is found that 0·624 grammes of silver are deposited in 20 minutes by the current which passes through voltameter and dynamometer. The torsion head was turned through 55° to balance the coils. What is the constant of the dynamometer?

12. What are the essential conditions of a good voltmeter and why?

13. In what respects does a voltmeter differ from an ammeter (*a*) in construction, (*b*) in use?

14. What would be the effect of connecting up a simple circuit with the voltmeter in series, and the ammeter in parallel with the source of E. M. F.?

15. What is meant by the specific resistance of a substance? How does it vary with temperature variations?

16. A wire 2 meters long and ·036 cms. diameter has a resistance of 1·78 ohms. What is the specific resistance of the material of the wire?

17. What is the resistance of a platinoid wire, 12 feet long and 0·025 inch in diameter? (See table of Specific Resistances.)

18. What length of No. 18 copper wire will have the same resistance as 1 mile of No. 18 silver wire?

19. What will be the joint resistance of the following wires in parallel:

100 yards of platinoid of diameter 0·01 inch;

100 metres of manganin of diameter 0·025 cm.;

1 mile of copper of diameter 0·01 inch?

20. A coil of copper wire whose *temperature coefficient* is 0·0038 has a resistance of 3·72 ohms at 0°C. What will be its resistance at a temperature of 15°C. and again at 100°C.?

21. Prove that the joint resistance R of a number of resistances r_1, r_2, and r_3, in parallel is given by

$$R = \frac{1}{\dfrac{1}{r_1} + \dfrac{1}{r_2} + \dfrac{1}{r_3}}.$$

22. Three resistances A, B, and C of 4, 6, and 12 ohms respectively are connected in parallel, and the common ends are joined to the terminals of a battery of 6·4 volts E. M. F. and 1 ohm internal resistance. What is the strength of the current in each resistance?

23. What will be the joint resistance of 250 yards of $\frac{7}{18}$ cable, 350 yards of $\frac{7}{16}$ cable and 1000 yards of $\frac{19}{18}$ cable in parallel? What must be the current strength in each cable when a D. P. of 10 volts exists across its ends?

24. An E.M.F. of 100 volts is applied to the ends of a resistance of 20 ohms. The current strength will be 5 amperes. By how much will this be altered when a voltmeter of 750 ohms is connected to the ends of the resistance?

What must be the least possible resistance of the voltmeter if the alteration is not to exceed 1 part in 1000?

25. What is the principle of the *shunt* as applied to high resistance galvanometers in certain measurements?

26. Explain why the method for determining high resistance could not be used for low resistance measurements, and vice versa.

27. Discuss the principle of the Wheatstone Bridge.

28. What is a *null* method, and what are its advantages over deflexional methods?

Will current variation affect the working of a Wheatstone Bridge? Will it affect the results of a low resistance measurement?

29. In a high resistance measurement using a universal shunt a deflexion of 135 was obtained with the $\frac{1}{100}$ shunt and 1 megohm resistance in the circuit. When an unknown resistance was substituted for the megohm a deflexion of 120 was obtained with the $\frac{1}{2}$ shunt. What was the unknown resistance? The combined resistance of galvanometer and shunt can be neglected.

30. If, in the above case, the galvanometer resistance was 10,000 ohms and the total resistance of the shunt (i.e. shunt 1) was 100,000 ohms, calculate the unknown resistance exactly.

31. Describe how a "shunted voltmeter" may be used as an ammeter.

32. A voltmeter of 3000 ohms resistance has a scale reading between 0 and 30 volts. If you had a current standard of $0 \cdot 1\ w$ resistance and you shunted the voltmeter across this, what "constant" would you have to multiply the reading by in order to get the amperes in the low resistance?

33. Describe a method of calibrating a voltmeter by means of a potentiometer.

34. How much work is being done in a circuit whose resistance is 6 ohms, when a current of 5 amps. is passed for 10 minutes? What is the rate of working (a) in watts, (b) foot lbs. per second?

35. If in the above case the resistance is reduced to 3 ohms, the E. M. F. remaining 30 volts, what will be the rate of working? How much heat would be developed in 1 hour?

36. What would be the cost of heating 1 gallon of water from 15° C. to 100° C. by electrical energy at $1\frac{1}{2}d.$ per B. O. T. unit?

37. What is the H. P. in the electrical circuit in which the energy boils 10 gallons of water in 30 minutes, the initial temperature of the water being 10° C.?

38. Describe how you measure the efficiency of an incandescent lamp.

39. Which type of lamp would be cheaper to run for 2000 hours under the following conditions?

	Life.	Watts per C.P.	First cost.	Cost per B.O.T. unit.	C.P.
A.	1000 hrs.	3·5	1s.	$4\frac{1}{2}d.$	32
B.	500 ,,	2·2	5s. 6d.	$4\frac{1}{2}d.$	32

40. Would it be good policy to overrun a carbon lamp at any stage of its life?

41. What are the points for and against "high-efficiency" lamps?

42. Describe briefly the Nernst incandescent lamp, and state the uses of its various parts.

43. Shew diagrammatically how you would arrange a Nernst lamp, a glow lamp, and an ordinary switch so that the glow lamp lights immediately the switch is closed, and is automatically put out when the Nernst filament glows.

44. Describe the working of a self-feeding arc lamp.

45. What is the function of a steadying resistance as used in connection with arc lamps?

46. A dynamo with an E.M.F. of 3000 volts is used to run 50 arc lamps in series. Each lamp has a candle-power of 400 and an efficiency of 1·5 watts per C.P. What is the current strength generated by the dynamo?

47. What are the differences between open and enclosed lamps, continuous and alternating current types?

48. 39 volts are absorbed at the crater of an arc, and the resistance of the arc circuit is 4 ohms. If a current of 8 amperes has to be passed through the lamp what must be the applied E.M.F.?

49. Discuss the relative merits of a "constant current" and a "constant E.M.F." system of electrical supply.

50. What are the advantages of the three-wire system over the two-wire?

51. A 20 H.P. dynamo has a difference of potential of 105 volts between the terminals. What will be the current in the circuit? If the circuit consists of forty 32 C.P. lamps, one hundred 16 C.P. lamps and twenty 100 C.P. lamps in parallel and the necessary feeders, find the drop in the feeders if the 100 C.P. lamps take 2 watts per candle, and the remainder take $3\frac{1}{2}$ watts per candle.

26—2

52. A pair of cables each 200 yards long supply 500 amperes to a distribution box, and the drop in the cables is 10 volts. What will be the current supplied when the drop is 3 volts? also find the drop when the current is 800 amperes.

53. The D.P. at the dynamo terminals is 220 volts and the distribution board is 500 feet away. What must be the cross-section of the cables if there are one hundred 60 watt, 110 volt lamps, supplied from the board, and the drop between dynamo and board is to be 2 per cent.?

54. What special method is commonly adopted in ship's wiring?

55. Explain, with sketches, what is meant by a "looping-in" system of wiring.

56. Apply the molecular theory to the explanation of the stages in the magnetisation of iron.

57. Define *Permeability*. How does the permeability of iron vary with the magnetising force? How does the magnetisation produced vary?

58. What is *Hysteresis*? Explain carefully what may be learned from the comparison of two hysteresis loops—one for iron and the other for steel.

59. A piece of iron is placed in a magnetic field of strength 6 c.g.s. units. The iron is 30 cms. long, and 1·5 sq. cms. area of cross-section. Its permeability under these conditions is 1000. What will be the magnetic moment of the iron?

60. In the above question, if the length of the iron was halved how would the answer be affected? If the area of cross-section be halved how would the answer be affected? Why?

61. Define *magneto-motive-force*; *reluctance* and *total* induction.

What is the length of a piece of iron whose area of cross-section is 5 sq. cms. and whose μ is 800, when a M.M.F. of 400 produces a total induction of 80,000?

62. What ampere-turns are necessary to magnetise an iron ring of mean circumference 100 cms. and area of cross-section 7·5 sq. cms. so that B shall be 12,000 lines per sq. cm.; the permeability of the iron being 950 for this value of B?

63. Work out the ampere-turns for a ring as in the preceding example but having an air gap of 0·3 cm. width.

64. What are the distinctions between self and mutual induction?

65. The magnetic field in a coil of wire consisting of 66 turns of wire is changed from 5,000 to 700,000 lines of force in 1·5 secs. What is the average E. M. F. induced in the coil?

66. Describe how a quantity of electricity may be measured by means of a ballistic galvanometer.

67. A coil is connected up to a ballistic galvanometer. A magnet is thrust into the coil (a) quickly, (b) not so quickly. Will there be any difference in the effects produced? If not, why not? Would the same answer hold good if an ordinary galvanometer were used?

68. A condenser of 0·3 m. f. d. capacity is charged so that a potential difference of 12·5 volts exists between its plates. What quantity of electricity will it yield on discharge?

69. Two condensers of 0·5 and 0·3 m. f. d. capacity are connected (a) in series, (b) in parallel. What is the combined capacity in each case?

70. A ballistic galvanometer needle gives a throw of 145 divisions when a condenser charged to 1·46 volts and of capacity 0·1 m. f. d. is discharged through it. Subsequently a coil of wire is connected to the galvanometer, the combined resistance being 5750 ohms. This coil is slipped off from the centre of a magnet. The coil has 100 turns of wire, and the deflexion given was 98 divisions. What is the pole strength of the magnet?

71. A "secondary coil" is connected to a ballistic galvanometer. This coil has 15 turns of wire and it is wound about a section of an iron ring. The primary coil consists of 110 turns of wire through which a current of 1·1 amperes flows. When the current is reversed the effect in the secondary causes a kick of the ballistic needle of 98 divisions. If the permeability of the iron is 700 and the combined resistance of galvanometer and secondary coil is 7000 ohms, what is the "quantity per division" for the ballistic galvanometer?

72. Describe the bar and yoke method of measuring permeability.

73. Why is the iron core of an armature laminated? Explain this fully.

74. Can you suggest any means of producing a unidirectional and steady E. M. F. by induction without commutation?

75. A single conductor is rotating in a uniform magnetic field at the rate of 500 revs. per minute. In each half revolution it cuts one million lines of force. What will be the average E. M. F., and what will be the instantaneous value of the E. M. F. in it when it has rotated through 75° from the zero position?

76. Give sketches of the connections for a ring armature of a four-pole alternator.

77. Sketch the connections for drum armature of a six-pole alternator. Shew poles and directions of E.M.F.s in coils.

78. Explain the principle of commutation with a two-part commutator.

79. Sketch the connection of drum armature—wave-wound—with 12 conductors and 6 commutator segments, for a two-pole machine.

80. Sketch the connections for a nine-part commutator, 18 conductor armature for a two-pole machine. Shew the pole pieces, the direction of rotation, and the direction of the E. M. F. in each conductor. Assume that the lead of the brushes is 10°, and shew the brushes in position.

81. An armature, wave-wound, has 50 conductors and 25 commutator bars. It rotates at 320 revs. per minute. If the machine is a two-pole machine and the E. M. F. yielded is 50 volts, what is the magnetic flux through the armature?

82. How do you account for the distortion of the magnetic field between two-pole pieces? Is the resultant field constant for variations of speed and armature current, assuming that the field magnets are permanent? How does it vary?

83. What can be learned from a characteristic curve?

84. Account for the difference between the total and the external characteristics of a series machine.

85. The special sphere of usefulness of a series machine lies in the running of a large number of arc lamps in series. Can you explain why this should be so?

86. Account fully for the behaviour of a shunt machine as its load is increased.

87. What is the coefficient of dispersion? How would you measure it?

88. What is the difference between the electrical and the mechanical efficiency of a motor?

89. In a brake test $(W - w) = 800$ grammes, the speed was 1100 revs. per minute and the pulley was 25 cms. diameter. The motor had an efficiency of 80 % and the applied E. M. F. was 100 volts. What was the current strength supplied?

90. Plot out the current curve for the circuit to which an impressed E. M. F. 100 virtual volts is supplied, the resistance being 3 ohms and the reactance being 4 ohms. Find the angle of lag.

91. A circuit is supplied with an alternating current, the impressed E.M.F. being 100 virtual volts. What will be the virtual amperes if the resistance is 1·5 ohms and the reactance is 3·5 ohms?

92. The reactance of a circuit is given by $2\pi f L$, where f is the frequency and L is the coefficient of self-induction of the circuit. What is L in that circuit in which an impressed E.M.F. of 220 virtual volts gives a virtual current of 100 amperes when the resistance is 1 ohm and the frequency is 80?

93. Plot out a sine curve to represent an alternating current. Then plot an active E.M.F. E when $E = 2 \times C$ at each instant. Then plot a back E.M.F. curve due to induction. This $E' = \frac{1}{4}E$ at each instant but 90° behind. From these plot out the *impressed* E.M.F. curve and *find the angle of lag*.

94. In the above question assuming that the *maximum value* of the active E.M.F. is 143 volts and the resistance is 2 ohms, determine the impressed E.M.F. (virtual), the virtual amperes, the impedance and the reactance.

95. Describe the principles and construction of a typical stationary transformer. What is a wattless current?

96. What types of practical instruments may be used to measure either continuous or alternating currents and E.M.F.s?

When used for alternating measurements what do they indicate—instantaneous, virtual, or average quantities? On what grounds do you base your answer?

97. If you had to charge a battery of 50 secondary cells in series, each cell with a capacity of 30 ampere-hours, what charging E.M.F. and current would you arrange for? If you did not know the capacity of the battery how would you determine the charging current?

98. Why is a shunt dynamo peculiarly suitable for the charging of accumulators?

99. What is the difference between a watt meter and a watt-hour meter?

100. An ampere-hour meter is graduated to read B.O.T. units on the assumption that it is to be used on a 100 volt supply. By mistake it is installed in a 230 volt supply. During a given period it records 747 units. How much would the company lose if they sent in a bill to the consumer for these 747 units at $4\frac{1}{2}d$. per unit?

DATA OF COMPOUND WOUND GENERATOR.

By Messrs SIEMENS Bros., Westminster.

Size 18 G.   Order No. 10232.   Machine No. 7603.

Output: 0/300 amps., 230/235 volts, 565/550 revs.

This machine is illustrated by Figs. 149 and 150. Fig. 151 is an open circuit characteristic curve shewing the relationship between the ampere-turns per pole and the resulting magnetic flux per pole (10^6 lines = 1 megaline).

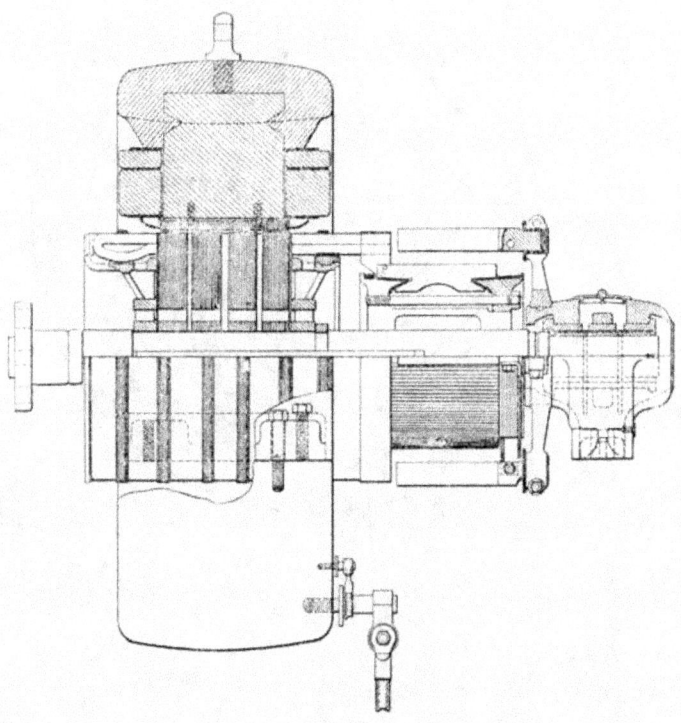

Fig. 149.

Armature:

No. and size of slots	42 – 13·1 mm. × 36 mm.
No. of coils	125.
Turns per coil	1.
Section of strip	14 mm. × 2·9 mm.
Winding connected	Wave.
Weight of copper	142 lbs.

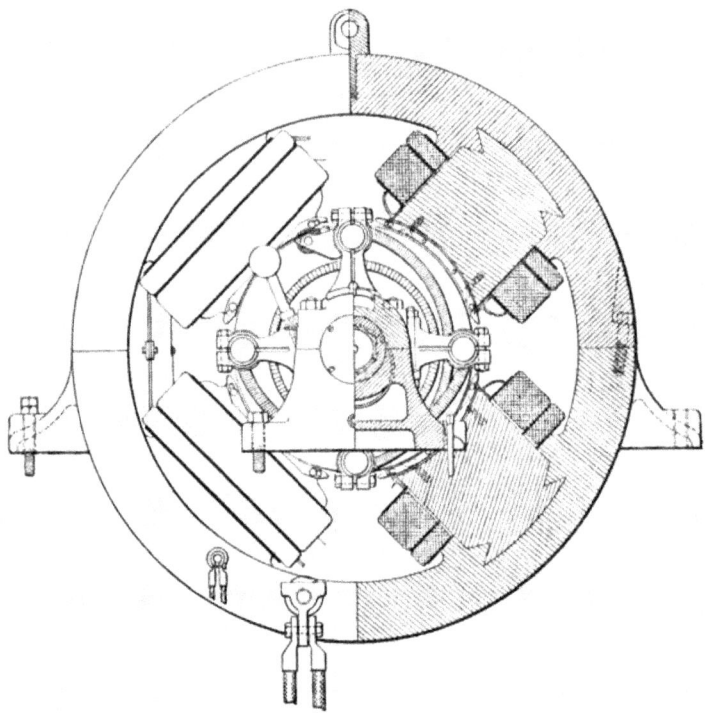

Fig. 150.

Commutator:

   125 parts.

Field:

  Shunt :

No. of coils	4.
Section of winding space	$3\frac{1}{2}'' \times 3\frac{3}{4}''$ deep.
Size of wire	1·7 mm. dia. bare.
Weight of wire	86 lbs. per coil.

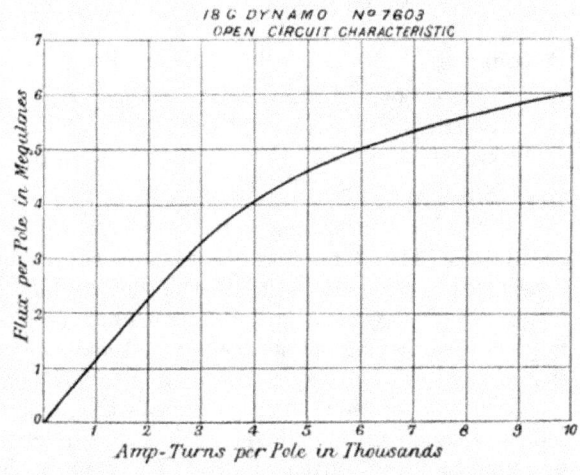

Fig. 151.

  Series :

No. of coils.	4.
Turns per coil	$8\frac{1}{2}$.
Section of copper	26 strips in parallel,
	14 mm. × ·6 mm. each.
Weight of copper	37 lbs. per coil.

DESCRIPTION OF 3-PHASE ALTERNATING CURRENT GENERATOR By MESSRS THE BRITISH ELECTRIC PLANT CO., LTD., ALLOA, SCOTLAND.

Output:

Voltage 6000.   Amperes 62·5.
Kilowatts 650.   cos ϕ = 0·85.
Revs. per min. 300.
Periodicity 50 $\sim$.
Number of poles 20.

Stator. The stator frame is built in two halves bolted together on horizontal centre line of the machine by fitted bolts. The frame is of high grade cast iron, designed to give freedom from sagging on the bottom half and to avoid vibration.

The stator core is built up of annealed high permeability soft steel stampings and these are insulated from each other to minimise eddy current losses. The core is thoroughly ventilated. The slots in the core are of the partially open type, the value of this lying in the fact that the coils can be entirely made by machine formers, and at the same time the advantages of the closed slot construction are in great measure retained. In the slots the coils are embedded in heavy split tubes of mica which insulate them thoroughly from the iron.

The coils are of special form and after winding on their formers are treated with insulating compound, pressed, and then stoved. After this they are covered on the parts outside the slots with special insulating material and tape. They are then again treated with compound and stoved.

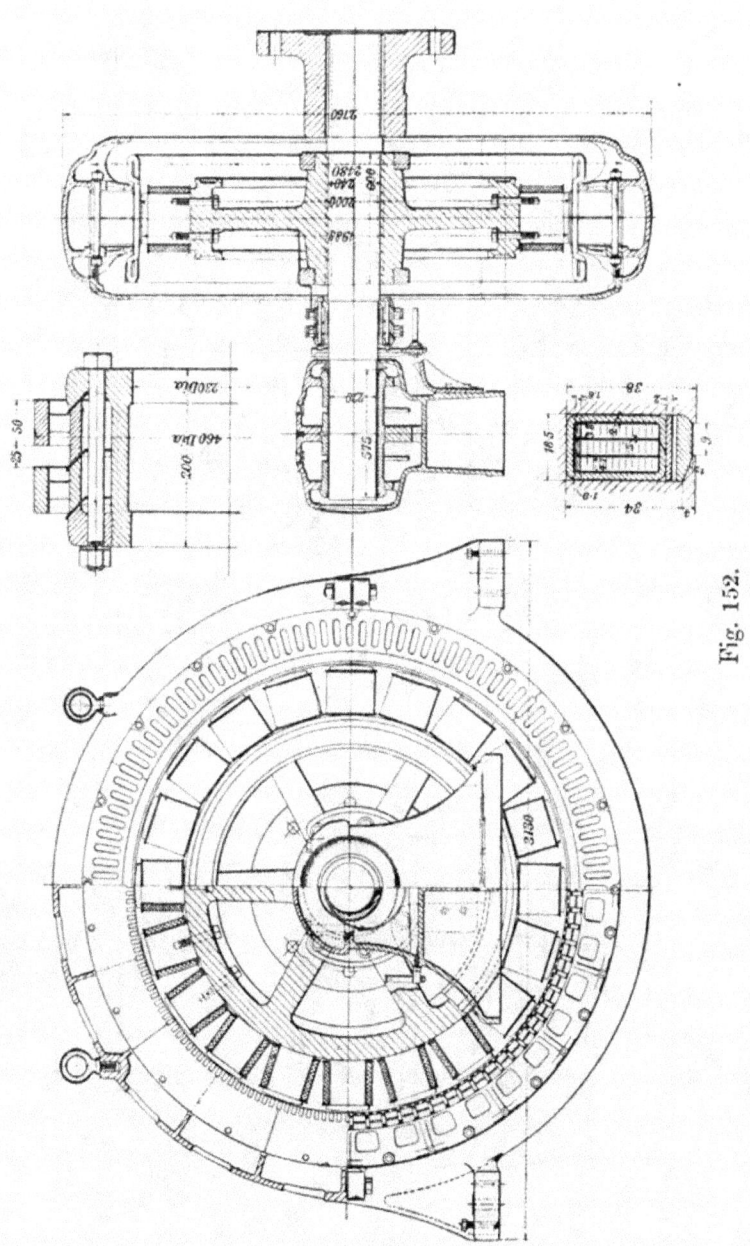

Fig. 152.

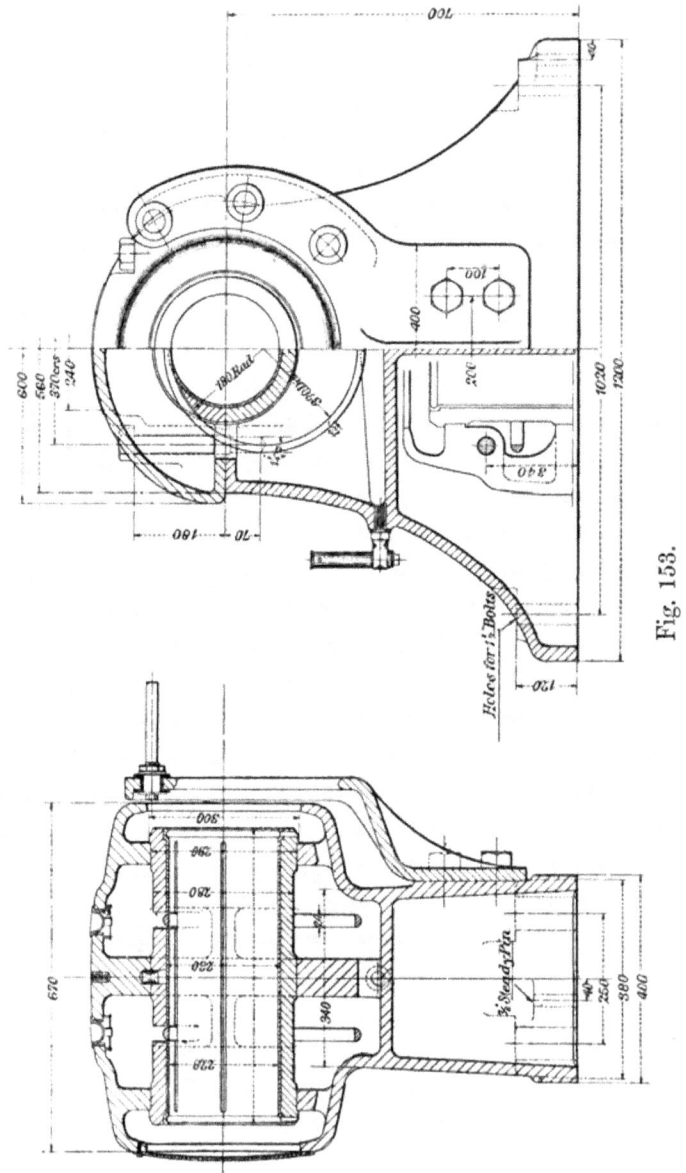

Fig. 153.

They are then thoroughly tested at double the maximum working voltage before being assembled on the machine. On the machine they are held in position in the slots by wedges of hard insulating material. The drawings and sections shew clearly the construction of the stator of this alternator.

The Rotor. The hub of the rotating field is a massive spider with heavy rim and is cast with the hub split on the centre line to minimise the stresses put on the casting when cooling. The hub is bored and slotted to fit the shaft and driving key and is held on the shaft by steel rings which are shrunk over turned projections on the split boss. The rim of the spider is of sufficient section to carry the magnetic lines from pole to pole and is accurately turned to accommodate the poles which are fastened to its outside periphery. The poles and pole facings are of cast steel, and the face of such shape as to give an air gap between the pole face and the inside of the stator discs of varying depth. At the centre of the pole the air gap is only 6 mm., whilst at the leading and trailing edges it is 10 mm. These small air gaps mean great accuracy in the machining and stamping operations.

The field coils are of heavy strip wound on edge in insulating spools, the adjacent turns having paper insulation between them.

The poles are fastened to the rim of the spider by fitted bolts and are readily removed from the rim if required.

The existing current for the field coils is provided by a small continuous current dynamo which can be direct coupled to the end of the rotor if a self contained generator is required.

The slip rings are of bronze, mounted on cast iron hubs secured to the shaft. The brush gear for supplying the current to the slip rings is carried on a bracket fastened to the end bearing. The brush holders are of a substantial form carrying carbon brushes of a lubricating nature and with wide range of feed so as to minimise the need of attention and the wearing of the slip rings.

The alternator shaft is of special open-hearth steel and of dimensions sufficient to prevent vibration, whilst the small distance between the bearings adds to its stiffness. On one end of the shaft a half coupling is carried for fastening to engine flywheel. In some cases the dynamo bearing is dispensed with and the generator is mounted between the cranks of a cross compound engine.

The bearings are of the self oiling ring lubricated type. The bearing bush is of cast iron lined with white metal. Inspection holes and oil sights and drain are provided.

Data.

Output.   See above.

Stator Core:

> External diameter = 2410 mm.
> Internal diameter = 2000 mm.
> Total axial length = 230 mm.
> Ventilating ducts, 1 in core 10 mm. wide.
>   ,,   ,,   2 outside core each 10 mm. wide.
> Nett axial length = 220 mm.
> Effective length of magnetic iron = 195 mm.
> Pole arc = 208 mm.
> Number of slots = 180.
> Size of slots = 18·5 mm. wide and 38 mm. deep.
> Opening of slot = 9 mm. (see dimensioned section).
> Thickness of core stampings = 0·5 mm.

Stator winding:

> Effective wires = $3 \times 780 = 2340$.
> Star connection.
> Section of wire—3 in parallel of $3·8 \times 1·6 = 3$ in parallel of 5·9 sq. mm. = 17·7 sq. mm. combined section.
> Covered size of wire $4·3 \times 2·1$ sq. mm.
> Mean length = 0·8 metre per wire per coil.
> Resistance per phase = $\frac{1}{3}$ of 2·1 ohms warm.

Air gap:

10 mm. on pole tips and 6 mm. at middle.

Poles:

Cast steel.

Section 225 mm. × 160 mm. Rounded ends.

Cross-sectional area = 300 sq. cms.

Thickness of pole face at middle = 25 mm.

Gross winding length = 200 mm.

Section of rim = 370 sq. cms. Cast iron.

Winding:

Section of wire = 3 mm. × 28 mm., bare strip.

Layers = 1. Turns = 52 per coil, wound on edge.

Mean length per turn 74 cms.

Excitation:

Amperes 205.

Volts 45.

Efficiency:

$\frac{5}{4}$ full load − 95 $^\circ/_\circ$.

Full load − 95 $^\circ/_\circ$.

$\frac{3}{4}$ full load − 94 $^\circ/_\circ$.

$\frac{1}{2}$ full load − 92·5 $^\circ/_\circ$.

$\frac{1}{4}$ full load − 88 $^\circ/_\circ$.

Drop in voltage from no load to full load with $\cos \phi = 0.85$ is 16 $^\circ/_\circ$.

This machine is illustrated by Figs. 152 and 153.

INDEX.

CAMBRIDGE: PRINTED BY JOHN CLAY, M.A. AT THE UNIVERSITY PRESS.

www.ingramcontent.com/pod-product-compliance
Lightning Source LLC
Chambersburg PA
CBHW081103170526
45165CB00008B/2307